or
Regul

Concepts in Radiation Cell Biology

CELL BIOLOGY: A Series of Monographs

G. M. Padilla, G. L. Whitson, and I. L. Cameron (editors). THE CELL CYCLE: *Gene–Enzyme Interactions*, 1969

A. M. Zimmerman (editor). HIGH PRESSURE EFFECTS ON CELLULAR PROCESSES, 1970

I. L. Cameron and J. D. Thrasher (editors). CELLULAR AND MOLECULAR RENEWAL IN THE MAMMALIAN BODY, 1971

I. L. Cameron, G. M. Padilla, and A. M. Zimmerman (editors). DEVELOPMENTAL ASPECTS OF THE CELL CYCLE, 1971

P. F. Smith. THE BIOLOGY OF MYCOPLASMAS, 1971

Gary L. Whitson. CONCEPTS IN RADIATION CELL BIOLOGY, 1972

Donald L. Hill. THE BIOCHEMISTRY AND PHYSIOLOGY OF *TETRAHYMENA*, 1972

In preparation

Kwang W. Jeon. THE BIOLOGY OF AMOEBA

Joseph A. Erwin (editor). LIPIDS AND BIOMEMBRANES OF EUKARYOTIC MICROORGANISMS

CONCEPTS IN RADIATION CELL BIOLOGY

Edited by GARY L. WHITSON

Department of Zoology and Institute of Radiation Biology
The University of Tennessee
Knoxville, Tennessee

1972

ACADEMIC PRESS New York and London

ACADEMIC PRESS, INC.
111 Fifth Avenue, New York, New York 10003

United Kingdom Edition published by
ACADEMIC PRESS, INC. (LONDON) LTD.
24/28 Oval Road, London NW1 7DD

LIBRARY OF CONGRESS CATALOG CARD NUMBER: 79-182619

PRINTED IN THE UNITED STATES OF AMERICA

This book is dedicated to Dr. Alexander Hollaender for introducing me to radiation biology

Contents

Chapter 3

The Effects of Ultraviolet Radiations on Bacteria

Barbara Ann Hamkalo

Chapter 4

Radiation-Induced Biochemical Changes in Protozoa

Gary L. Whitson

Chapter 5

Techniques for the Analysis of Radiation-Induced Mitotic Delay in Synchronously Dividing Sea Urchin Eggs

Ronald C. Rustad

Chapter 6

The Effects of Radiation on Mammalian Cells

Carl J. Wust, W. Stuart Riggsby, and Gary L. Whitson

Chapter 7

Ionizing Radiation Effects on Higher Plants

Alan H. Haber

Chapter 8

Photodynamic Action of Laser Light on Cells

I. L. Cameron, A. L. Burton, and C. W. Hiatt

List of Contributors

Numbers in parentheses indicate the pages on which the authors' contributions begin.

A. L. BURTON (245), Departments of Anatomy and Bioengineering, The University of Texas Medical School at San Antonio, San Antonio, Texas

I. L. CAMERON (245), Departments of Anatomy and Bioengineering, The University of Texas Medical School at San Antonio, San Antonio, Texas

ALAN H. HABER (231), Biology Division, Oak Ridge National Laboratory, Oak Ridge, Tennessee and Department of Botany, The University of Tennessee, Knoxville, Tennessee

BARBARA ANN HAMKALO (91), Biology Division, Oak Ridge National Laboratory, Oak Ridge, Tennessee

C. W. HIATT (57, 245), Department of Bioengineering, The University of Texas Medical School at San Antonio, San Antonio, Texas

RONALD O. RAHN (1), Biology Division, Oak Ridge National Laboratory, Oak Ridge, Tennessee

W. STUART RIGGSBY (183), Department of Microbiology, The University of Tennessee, Knoxville, Tennessee

RONALD C. RUSTAD (153), Department of Radiology and Developmental Biology, Center Case Western Reserve University, Cleveland, Ohio

GARY L. WHITSON (123, 183), Department of Zoology and Institute of Radiation Biology, The University of Tennessee, Knoxville, Tennessee

CARL J. WUST (183), Department of Microbiology, The University of Tennessee, Knoxville, Tennessee

Preface

One of the most difficult tasks in compiling and editing a technical book such as this is to present, without loss of continuity, a central theme that encompasses and unifies the current concepts of the field. The field of cell biology itself is very broad, and cellular phenomena are often as numerous and varied as the kinds of cells that exist either as unicellular organisms or as tissue elements of plants and animals. Therefore, any special subject area within the realm of cell biology, such as the effects of radiations on cells, could be large and diverse; and it is.

It has not been possible to review the entire field of radiation cell biology. Rather, in this book, current concepts of various cellular radiobiological phenomena in selected cell types are surveyed. Since most books on this subject are of a review nature, the decision was made to present some important techniques that offer students and investigators the necessary information for further experimentation in radiation cell biology.

Although not readily apparent in each chapter, the general theme resides in the underlying macromolecular basis for cellular changes in irradiated cells. The first few chapters deal mainly with the effects of nonionizing radiations ranging from ultraviolet to visible light. The remaining chapters, except the last one, which presents the use of laser light in cellular studies, deal mainly with ionizing radiations. The evidence thus far obtained from ultraviolet studies implicate DNA as the main target macromolecule responsible for such radiation injury as division delays, delayed DNA replication, and lethality. On the other hand, photodynamic effects also favor proteins as important target molecules for radiation injury in cells. Recent evidence on ionizing radiation also favors DNA as the main target molecule responsible for cellular radiation injury, but less is known about specific biochemical changes and repair processes in cells exposed to this type of radiation.

This book is aimed primarily at the level of advanced students in radiation biology. It begins with physical–chemical studies on ultraviolet-irradiated DNA. The remaining chapters, beginning with viruses, are organized in an increasing order of complexity of cell types, including cells of higher plants.

GARY L. WHITSON

This photograph of Dr. Hollaender was taken at Dartmouth College in August, 1968, at the Fifth International Congress on Photobiology where he was awarded the Finsen Medal, which he is holding in his left hand. The medal was awarded to Dr. Hollaender "for fundamental contributions in the early development of photobiology, in particular radiation genetics." Permission to reproduce this photograph was granted through the courtesy of Per Hjortdahl of the Dartmouth Medical School, Hanover, New Hampshire.

Chapter 1
Ultraviolet Irradiation of DNA

RONALD O. RAHN

I. Introduction

A principal aim in modern radiation biology is to interpret the effects of radiation on living systems in terms of the changes observed in cellular components and functions. A typical experiment consists of irradiating a cell under well-defined conditions and then measuring the effect of the radiation on one of several biological factors, such as the ability to form colonies, the rate of cell growth and division, and the rate of mutation. One is interested in how these gross biological effects are correlated with changes in the rate of cellular functions such as the synthesis of DNA, RNA, or protein. Such correlations are often difficult to make, partly because of the myriad of metabolic control mechanisms that may be in various stages of breakdown and partly because of enzymatic repair of the damaged DNA. In recent years a strong effort has been made to approach this problem at the molecular level and ask: What are the chemical changes, particularly in DNA, that occur during and following the irradiation of a cell and how can these events affect the various cell functions?

The various events that follow the interaction of a cell with radiation are diagramed in Fig. 1. Designated in the figure are three broad areas of research in radiation biology in which one can carry out experiments and organize data: They are the physicochemical, biochemical, and biological aspects of the interaction between radiation and the cell. Linking these areas is a challenging problem, and thus far only a limited degree of success has been achieved.

In this chapter we shall concern ourselves with the contents of the first box in Fig. 1. In particular, we shall deal only with the effects of UV* on

* Abbreviations and symbols: BrUra, 5-bromouracil; PO-T, deamination product of the cytosine–thymine adduct, 6-4′-[pyrimidin-2′-one]-thymine; UV, ultraviolet radiation; $\widehat{TT}_1$, *cis-syn* thymine–thymine cyclobutane dimer; $\widehat{TT}_2$, *trans-syn* thymine–thymine cyclobutane dimer; $\widehat{CC}$, cytosine–cytosine cyclobutane dimer; $\widehat{CT}$, cytosine–thymine cyclobutane dimer; Y, unspecified pyrimidine nucleoside; N, unspecified

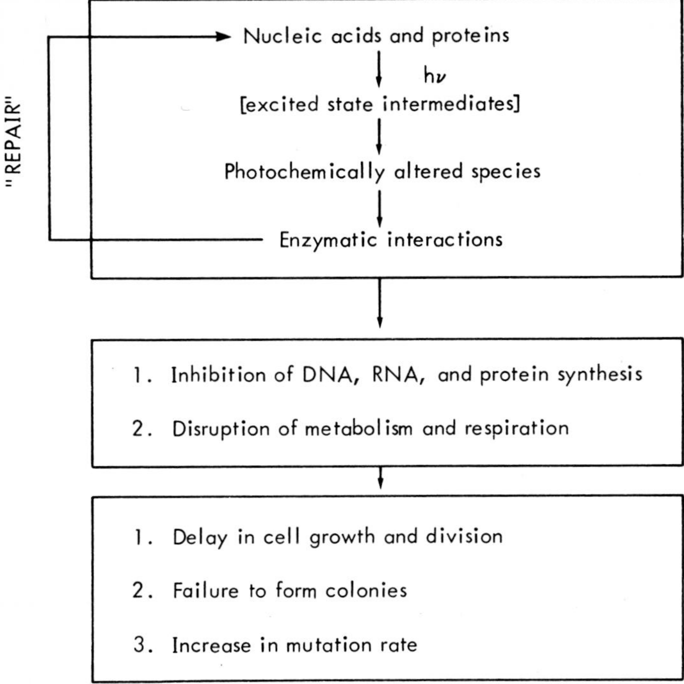

Fig. 1. Schematic representation of events following the ultraviolet irradiation of a cell.

the properties of isolated DNA molecules. We shall try to answer the question: What kinds of damage are produced in DNA by UV-irradiation, and how does this damage affect the interactions between DNA and enzymes? The remaining chapters will deal with the effects of radiation on intact biological systems.

The reason for focusing on the photochemistry of DNA and not on that of RNA or protein is that DNA occupies the most important position in the cell's machinery, as judged by the order of synthesis:

$$\text{DNA} \rightarrow \text{RNA} \rightarrow \text{protein}$$

Hence, alteration of DNA by UV affects those processes that are dependent upon the integrity of the DNA template and can thus lead to lethal or mutagenic effects. Alteration of the RNA or protein component of a cell would not have as devastating an effect on the functions of a cell because extra copies of these molecules are available.

We will pay particular attention to environmental factors such as

---

nucleoside; poly(dA-dT), alternating copolymer of A and T; poly(dA)·poly(dT), association of poly(dA) with poly(dT); hyphens within base sequences indicate a phosphoric diester in 3′-5′ linkage.

temperature and relative humidity, which have been found to strongly influence the rate of formation of DNA photoproducts. An understanding of these factors should help the researcher design experiments in which the physical environment of the DNA in a cell is altered and the resulting change in the photobiological effect correlated with changes in the DNA photochemistry. For example, cells are normally irradiated at room temperature in neutral buffer solutions. By varying these physical conditions, one may be able to observe a new dependence of the biological effect on dose and may be able to correlate this change with a concomitant change in the number and nature of photoproducts formed. When irradiating cells, the freedom to vary the irradiation conditions is somewhat limited because the cells may lose their viability if subjected to extreme changes in their chemical and physical environment. However, photobiological investigations have been carried out with bacterial spores (Donnelan *et al.*, 1968), vegetative cells (K. Kaplan, 1955; Smith and O'Leary, 1967; Webb, 1965), and bacterial viruses (Fluke, 1956; Hill and Rossi, 1954; Levine and Cox, 1963) subjected to extremes in temperature and relative humidity during irradiation. In some of these experiments rather great changes in the production of photoproducts were observed and these were correlated with the observed biological inactivation curves. The biological activity of transforming DNA has also been measured after irradiation over a wide range of temperatures (Rahn *et al.*, 1969).

This chapter is not intended to be a review of nucleic acid photochemistry and photobiology. For recent reviews of this subject, see J. K. Setlow (1966a) and R. B. Setlow (1968a). One of the early attempts to review nucleic acid photochemistry was made by McLaren and Shugar (1964); their book still contains a wealth of information and ideas not presented elsewhere. A recent book on molecular photobiology has been written by Smith and Hanawalt (1969). A review of the photochemistry of nucleic acid derivatives from the viewpoint of the chemist has been presented by Burr (1968).

II. Photophysics

A. Light Sources

We shall deal only briefly with the techniques used to incorporate energy from some outside source into the biological system under consideration. More detail on this problem is provided by Jagger (1967) and Johns (1968), and the reader is advised to consult these references for additional information on experimental details. In addition, Chapter 2 in this book describes in detail some of the techniques used in irradiating viruses. These techniques are similar to those used for irradiation of other systems, such as bacteria and protozoa.

There are several different kinds of lamps available for obtaining ultraviolet radiation. In most experiments, monochromatic radiation is essential. A 15-W, low-pressure mercury lamp is widely used in photobiology. It is convenient to use and emits most of its energy (~85%) at 254 nm. Hence, no filters or monochromators are necessary to obtain radiation predominantly at this wavelength, which is close to the absorbance maximum of DNA (257 nm). However, if one wishes to irradiate at different wavelengths, as when measuring an action spectrum, then a source with a broader spectral output must be used in conjunction with a band-selecting device. Two possible sources are the high-pressure mercury lamp and the xenon lamp. The relative spectral output of these is given in Fig. 2. The output of the high-pressure mercury lamp is concentrated at certain wavelengths, the principal lines. A band-selecting device such as an interference filter or a monochromator is used to select the line of interest. An advantage of the xenon lamp is a spectral output that is continuous in the ultraviolet so that one is not restricted to using only certain wavelengths, as with the high-pressure mercury lamp. Hence, a xenon lamp is better suited for obtaining a high-resolution action spectrum.

Muel and Malpiece (1969) have described a set of solution filters that

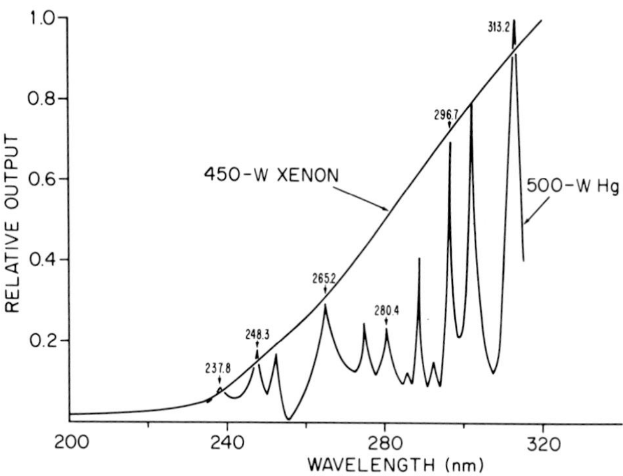

Fig. 2. Relative output in the ultraviolet region of a 450-W xenon lamp compared with that of a 500-W, high-pressure mercury lamp. The numbers indicate the wavelengths of the principal mercury lines commonly used for irradiation.

will isolate various portions of the UV spectrum. These filters are inexpensive and permit one to carry out irradiations at various wavelengths without using expensive monochromators or thin-film interference filters. One of their shortcomings, however, is their photolability, which necessitates changing the solutions constantly during any prolonged irradiation.

B. Absorption

In order for a photochemical reaction to occur, a photon must be absorbed by (or its energy transferred to) the system in which the reaction takes place. The probability that an absorbed photon will give rise to a photochemical reaction is called the quantum yield (Φ) of the reaction and is equivalent to the ratio of the number of molecules altered to the number of photons absorbed. In a cell there are two classes of macromolecules mainly responsible for the absorption of UV—proteins and nucleic acids. In the region 240 to 280 nm, the nucleic acids, in small cells, are the most important physical absorbers of light. In this wavelength region, nucleic acid absorption is 10 to 20 times greater by weight than that of protein. The absorbance (A) at some chosen wavelength of a solution of DNA in a 1-cm cell may be written as $A = c \cdot \epsilon$, where c is the concentration in moles per liter and ϵ is the molar absorptivity or extinction coefficient. Extinction coefficients for polynucleotides are usually expressed as $\epsilon(P)$, the molar extinction coefficient per nucleotide (phosphate). Generally, all DNA's, regardless of their G + C content, show values of $\epsilon(P)$ at 260 nm that vary from 6600 to 7000. The absorbance of DNA can be considered to be the sum of the absorbances of the individual nucleosides, with some modification due to hypochromicity. The hypochromicity or decrease in molar extinction upon incorporation of the nucleosides into native DNA is approximately 30% and arises from electronic interactions between the stacked bases.

The action spectrum of a system corresponds to the wavelength dependence of some light-induced change in the system. It is determined by measuring, at various wavelengths of irradiation, the number of incident photons needed to produce an observed change in the system. For optically dilute samples (transmission approx. 90%), the fraction of incident energy absorbed by a molecule at a given wavelength is nearly proportional to the extinction coefficient of the molecule at that wavelength. Hence, the number of incident photons needed at some wavelength to achieve a fixed percentage of change in a molecule will be inversely proportional to the extinction coefficient of the molecule at that wavelength, provided the quantum yield is independent of wavelength. Therefore, a plot of the inverse number of photons needed at each wavelength to achieve a fixed percentage of change should resemble the absorption spectrum of the molecule. A good description of how to obtain an excitation spectrum is given by Jagger (1967).

In many biological systems, the action spectrum for UV-induced biological changes has been found to resemble the absorbance spectrum of nucleic acids. The action spectrum for the inactivation of a culture of *Escherichia coli* was obtained by Gates (1930) and is Fig. 1 in Chapter 3. This spectrum was one of the first indications that nucleic acids are the primary target for the lethal effect of UV. A discussion of UV action spectra with particular emphasis on mutation is given by Giese (1968).

C. Excited States

The absorption of a photon promotes a molecule to an excited electronic state. A molecule in such a state is often more chemically reactive than when in a ground state. As indicated in Fig. 3, a molecule in an excited state has a choice of pathways by which it can return to the original

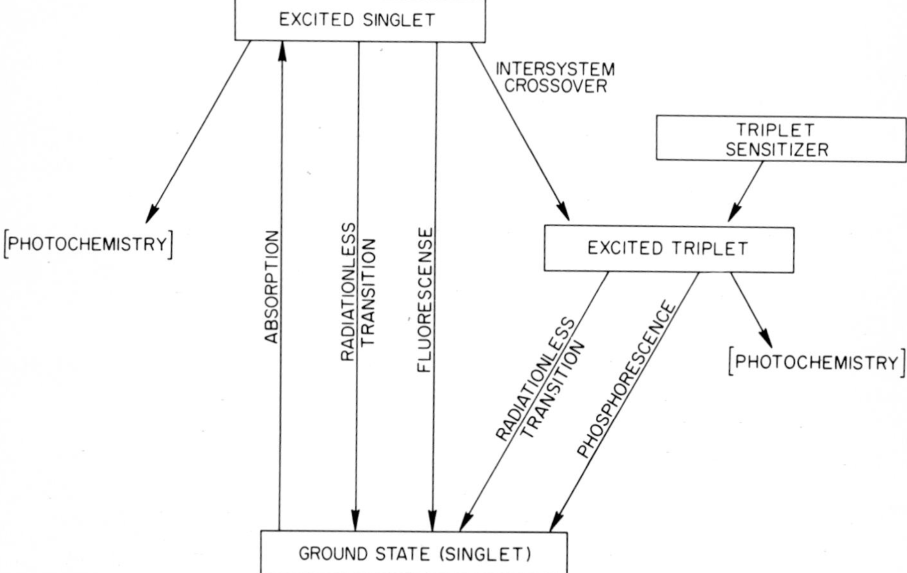

Fig. 3. Diagram of the electronic energy levels of a molecule indicating the various types of transitions between the lowest-lying excited states and the ground state.

ground state, or it can undergo a chemical transformation to form a new species. That is

$$A \overset{h\nu}{\rightleftarrows} A^* + B \rightarrow C$$

where A^* is an excited state of A. Usually the energy lost in going back to the ground state is given up in the form of heat, although light can be emitted in the form of fluorescence or phosphorescence. Normally, however, the emission from nucleic acids is very weak and can be observed only at temperatures below 100°K.

With purine and pyrimidine bases, the excited state initially formed after the absorption of a photon is an excited singlet state and corresponds to the excitation of an electron from a bonding π orbital to a higher-lying π^* orbital. The excited electron in the π^* orbital can be either paired (singlet state) or unpaired (triplet state) with the single electron remaining in the π orbital. The lower-lying triplet state is reached from the singlet state by a process known as intersystem crossing. Because of the "spin forbiddenness" of the transition between the triplet state and the ground state, the lifetime of the triplet state is many orders of magnitude longer than that of the singlet state.

Once a molecule absorbs a photon, the excitation energy can be transferred to some other molecule in the system provided the acceptor molecule has a lower-lying set of energy levels than the donor. Energy transfer between a donor D and an acceptor A can be written

$$D^* + A \rightarrow D + A^*$$

Transfer can be either at the singlet level or the triplet level. Singlet transfer is generally thought to occur via a resonance dipole–dipole energy transfer mechanism (Förster, 1959). The probability of transfer between two molecules is related to the dipole strengths, the separation distance, the relative orientation of the dipoles, and the overlap of the donor emission and the acceptor absorbance. Singlet transfer may occur over distances of up to 40 Å (Förster, 1959) under favorable conditions. In DNA, however, long-range, singlet–singlet transfer between the bases is not likely, in part because certain bases or base pairs may form energy sinks (Gueron and Shulman, 1968). Singlet transfer, on the other hand, may occur from the bases to bound dyes such as proflavin and acridine orange, which have low-lying excited singlet levels (B. M. Sutherland and Sutherland, 1969). Also, mercuration of the bases in DNA creates some bases with low-lying excited electronic levels, which presumably act as energy traps for the thymine energy (Rahn *et al.*, 1970).

Triplet transfer proceeds by an exchange interaction (Dexter, 1953), and it is necessary for the molecules undergoing energy exchange to be close enough for orbital overlap to occur. Adjacent bases in DNA are sufficiently close to allow triplet transfer; such a mechanism may explain the long-range quenching of low-temperature phosphorescence of DNA by bound paramagnetic metal ions (Bersohn and Isenberg, 1964). Triplet transfer from an exogenous molecule to DNA is also possible. In a later section, the photochemical processes that follow the transfer of triplet energy from an excited donor molecule, such as acetone or acetophenone, to DNA are discussed. This type of transfer takes place in solution when the excited donor molecule comes close enough to a base for exchange to occur.

In DNA, a base in an excited singlet state can interact electronically with its ground-state neighbor to form an excited-state dimer, or exciplex (Eisinger, 1968). In the exciplex, the two molecules move closer to one another and the excitation energy is distributed over both molecules. The exciplex singlet energy level is lower than that of the initially excited monomer, and in many polynucleotides, including DNA, the fluorescence is associated with this low-lying exciplex level. Hence, exciplex fluorescence is broad and shifted to the red relative to the fluorescence of the isolated bases. The exciplex plays an important role in DNA photochemistry since it is a possible excited-state precursor for the formation of pyrimidine dimers (Eisinger and Lamola, 1967).

III. Identification of Photoproducts

A. Absorbance Measurements

Given an irradiated sample of DNA, the problem is to determine what kind of photoproducts are present and how many of each there are. In some cases absorbance measurements on the intact DNA have been useful in determining the presence of certain photoproducts. R. B. Setlow and Carrier (1963), for example, showed that the formation of cytosine and thymine dimers (Section IV,B) in DNA results in a decrease in absorbance at 260 nm because of the saturation of the pyrimidine 5–6 double bond. Photoreversal of these dimers with short-wavelength irradiation leads to an increase in the 260 nm absorbance. Such behavior is characteristic of cyclobutane dimers. Similarly, another photoproduct, the cytosine hydrate (Section IV,E), is heat unstable, so heat-reversible changes in absorbance can be related to the presence of such hydrates (R. B. Setlow and Carrier, 1963). Other photoproducts have unique absorbance properties and are easy to detect optically. The thymine–

cytosine adduct (Section IV,D), for example, contains a pyrimidin-2′-one ring with an absorbance maximum at 315 nm. Absorbance measurements have also been used to measure the formation of UV-induced cross-links in DNA (Rahn *et al.*, 1969). Since cross-links prevent complete strand separation of DNA upon heating, the amount the absorbance decreases upon cooling, due to reassociation of the strands, is a measure of cross-link formation.

One complication associated with trying to correlate absorbance changes with photoproduct formation is the concomitant increase in the DNA absorbance brought about by changes in the degree of hypochromicity. Photoproducts cause disruption of the hydrogen bonding and base stacking, which results in locally denatured regions and a subsequent loss of hypochromicity in these regions. It is necessary, therefore, to consider this loss when interpreting the absorbance properties of irradiated DNA.

There is another factor that limits the usefulness of absorbance measurements for analyzing irradiated DNA. The percentage of absorbance change in DNA after UV-irradiation is small because the majority of bases do not react photochemically. Even at very high doses of radiation, the maximum decrease in absorbance is about 20%. Another disadvantage of this method is that optically detectable amounts of DNA are required for analysis.

The advantages of using absorbance measurements to analyze the photoproducts formed in DNA are (1) intact DNA can be examined, thus eliminating hydrolysis, and (2) radioactively labeled DNA is not needed.

B. Chromatographic Detection of Labeled Photoproducts

DNA photochemistry has been investigated primarily with liquid-phase chromatography, to separate the photoproducts, and radioactive counting techniques, to determine the relative concentration of each photoproduct. This method of analysis differs from absorbance measurements in that the DNA has to be radioactively labeled and the analysis is made on DNA that has been hydrolyzed to monomer units plus photoproducts. A more complete description of the chromatographic analysis of irradiated DNA is given by Carrier and Setlow (1971).

1. *Labeling the DNA*

Normally, either carbon-14(^{14}C), with a half-life of 5700 years, or tritium (^{3}H), with a half-life of 12.5 years, is used as a radioactive label.

Radioactive nucleosides are readily available and can be added to the growth medium to obtain organisms with labeled DNA. The DNA is then isolated from the cells in a normal fashion. It is relatively simple to prepare thymine-labeled DNA with high specific activity from many microorganisms, since there are numerous thymine-requiring mutants. It is more difficult, however, to obtain thymine-labeled DNA from organisms without a specific requirement for thymine or to obtain DNA labeled at one of the other bases. Nevertheless, photoproducts have been measured in a wide variety of organisms and in cultured cells whose DNA has been labeled with radioactive thymidine simply by adding it to the growth medium. Some systems, such as yeast, however, will not take up labeled nucleosides, so photochemical studies in these systems are, for the most part, lacking.

Labeled DNA suitable for photochemical studies can also be obtained by means of the calf-thymus polymerase system (Bollum, 1960). The base composition of DNA obtained in this way will reflect the (A + T)/(G + C) ratio of the template DNA. R. B. Setlow and Carrier (1966) prepared ^{3}H-cytosine-labeled DNA by mixing the appropriate deoxyribonucleoside triphosphates with calf-thymus DNA polymerase in the presence of DNA template. The resulting DNA's varied widely in their nearest-neighbor frequencies, depending upon the choice of template DNA.

2. *Hydrolysis*

DNA can be hydrolyzed by keeping it at 175°C for 30 min in 98% formic acid or at 155°C for 60 min in trifluoroacetic acid. This treatment converts DNA into individual purine and pyrimidine bases plus acid-stable photoproducts, and destroys photoproducts that are labile to heat and acid. In some cases a transformation of the initial photoproduct to another form may take place. An example is the deamination at high temperatures of cytidine when its C-5 to C-6 double bond has been saturated (R. B. Setlow and Carrier, 1966). Hence, there are difficulties inherent in this technique that complicate the task of relating the information obtained from hydrolyzates to the state of the irradiated DNA prior to hydrolysis.

It is possible to use enzymatic hydrolysis to cleave the phosphodiester bonds of single-stranded DNA and thus obtain a mixture of the individual nucleotides. The problem in applying this technique to the degradation of irradiated DNA is that the nucleases are not capable of cleaving the phosphodiester bond between a pyrimidine dimer and a neighboring base (R. B. Setlow *et al.*, 1964). The chromatography is thus complicated by an additional monomer unit being attached to every dimer.

3. *Analysis*

a. Paper Chromatography. The most commonly used technique for separating nucleic acid photoproducts is paper chromatography (descending). The hydrolyzate of the irradiated DNA is spotted at the origin of the chromatogram, which is normally composed of Whatman No. 1 filter paper. The spot is kept as small as possible in order to obtain maximum resolution of the photoproducts. Not more than 200 μg of DNA should be contained in each spot. The chromatogram is then placed in contact with the solvent. A list of solvents commonly used for chromatographic analysis of nucleic acid photoproducts is given in Table I. In 12–18 hr the solvent front will have traveled about 35 cm, and the photoproducts will be distributed somewhere between the origin and the solvent front. The chromatogram is either counted intact with a strip scanner, or the paper is cut into strips and the radioactivity in each strip counted separately.

The mobility of a photoproduct on paper saturated with a given solvent depends upon its solubility in that solvent. The distance a photoproduct moves from its origin relative to the solvent front is called an R_f value; this value usually changes with each different solvent (Table I). Sometimes two different photoproducts will have the same R_f value in a given solvent; consequently, in identifying a photoproduct for the first time, it is important to determine the R_f value in as many solvents as possible.

Shown in Fig. 4 is the distribution of radioactivity along chromatograms of hydrolyzates of irradiated DNA labeled either with radioactive thymine or cytosine. Both samples received the same dose of UV. Deamination of the cytosine-containing photoproducts during hydrolysis results in the conversion of C to U, so cytosine-containing photoproducts appear as uracil products. The thymine-labeled DNA has three peaks, corresponding to unreacted thymine (T) plus two pyrimidine dimers labeled $\widehat{\mathrm{TT}}$ and $\widehat{\mathrm{UT}}$. The exact nature of these dimers will be discussed in Section IV. The cytosine-labeled DNA contains the pyrimidine dimer $\widehat{\mathrm{UU}}$ derived from $\widehat{\mathrm{CC}}$, in addition to $\widehat{\mathrm{UT}}$. The percentage of activity obtained as the $\widehat{\mathrm{UT}}$ dimer is nearly the same for both samples because the amounts of cytosine and thymine present in *E. coli* DNA are nearly the same.

A precise way of characterizing an unknown photoproduct, P_1, which is thought on the basis of its R_f value to be the same as P_2, a previously identified photoproduct, is to mix the two differently labeled photo-

TABLE I

R_f Values of DNA Photoproducts for a Variety of Solvents Used in Either Paper or Thin-Layer Chromatography; Also Included Are the Tube Numbers Used in Locating Photoproducts Separated by Dowex Column Chromatography

Photoproduct	R_f values of solvent system[a]					Tube No.[b]
	1	2	3	4	5	
Thymine (T)	0.48[c]	0.60–0.63[h]		0.46[c]	0.65[b]	60[c]
Thymine–thymine dimer ($\widehat{TT}_1$)	0.1[c]–0.12[d]	0.24–0.31[h]	0.49[c]	0.66[c]	0.55[k]	26[c]
Thymine–thymine dimer ($\widehat{TT}_2$)	0.18[d]–0.24[d]	0.43[i]	0.62[c]	—	0.65[i]	
Uracil (U)	0.35[e]	0.46–0.48[h]				
Uracil–uracil dimer ($\widehat{UU}$)	0.04[e]	0.09–0.11[f]				26[l]
Uracil–thymine dimer ($\widehat{UT}$)	0.1–0.12[f]	0.19–0.22[f]			0.41[k]	26[l]
Spore photoproduct	~0.14[g]	0.38[j]				
6-4′-[Pyrimidin-2′-one]-thymine (PO-T)		0.24–0.31[k]			0.29[k]	
Cytosine (C)	0.19[c]	0.29–0.36[h]		0.66[c]		20[c]
Dihydrothymine (TH_2)	0.45[c]	0.64[c]		0.55[c]		17[c]

[a] Solvent systems: 1. *n*-Butanol–water (86:14); 2. *n*-Butanol–water–acetic acid (80:30:12); 3. Isopropanol–ammonia–water (7:1:2); 4. Saturated ammonium sulfate–1 *M* sodium acetate–isopropanol (40:9:1); 5. *t*-Butanol–methyl ethyl ketone–water–ammonia (40:30:20:10, v/v).

[b] For use with Dowex 1-X2 column ammonium formate gradient.

[c] Yamane *et al.* (1967).

[d] Weinblum and Johns (1966).

[e] Greenstock and Johns (1968).

[f] R. B. Setlow and Carrier (1966).

[g] Donnellan (1971).

[h] Smith (1963).

[i] Ben-Hur and Ben-Ishai (1968).

[j] Donnellan and Setlow (1965).

[k] Varghese and Wang (1968).

[l] Lamola (1969).

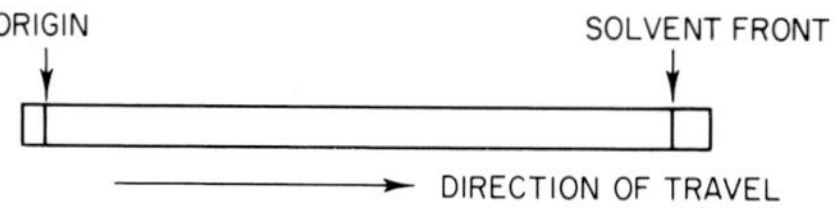

products together and chromatograph them simultaneously. One photoproduct, for example, may be labeled with ^{14}C and the other with ^{3}H. The ^{14}C and ^{3}H activities are counted in separate channels of a scintillation counter and plotted as a function of the distance from the origin. The two curves, if the photoproducts are the same, should be superimposable. One then says that P_1 is chromatographically indistinguishable from P_2.

When working with very small amounts of photoproducts ($<1\%$

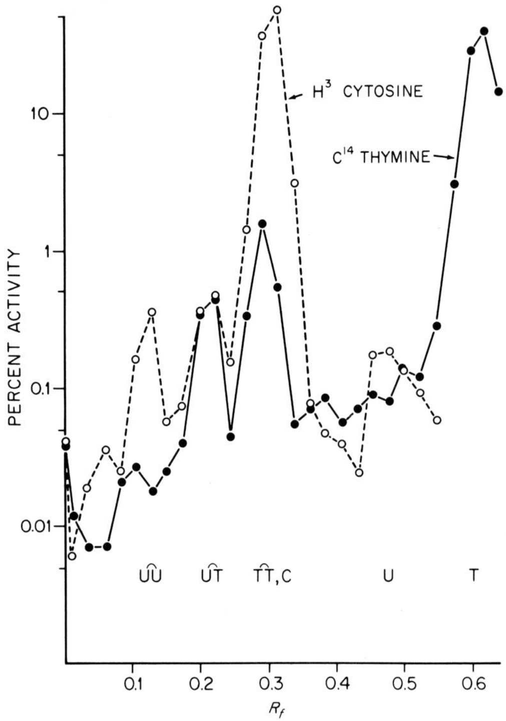

Fig. 4. Paper chromatogram developed with *n*-butanol–acetic acid–water of a hydrolyzate of irradiated DNA labeled with either ^{3}H-cystosine or ^{14}C-thymine. Both DNA's received the same dose (280 nm, 1×10^{4} ergs·mm^{-2}). (From R. B. Setlow and Carrier, 1966.)

yield), it may be necessary to use two-dimensional chromatography to accurately measure the yield. This technique gets rid of the thymine streaking, which may obscure low levels of photoproducts. Essentially one runs a chromatogram along one edge of the paper and then rotates the paper 90° and runs a second chromatogram at right angles to the first one. The second chromatogram may or may not be run in a different solvent. The chromatogram in the second dimension is then analyzed with the first-dimensional chromatogram as the origin.

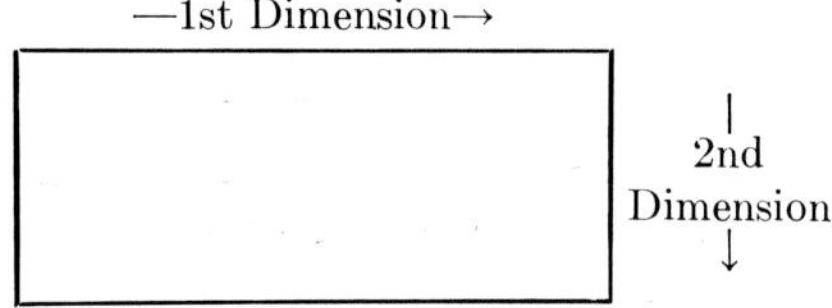

b. Thin-Layer Chromatography. Nucleic acid photoproducts can also be separated by thin-layer chromatography (B. M. Sutherland and Sutherland, 1969; Greenstock and Johns, 1968). Basically, this technique is very similar to paper chromatography except the support matrix consists of cellulose plates on a plastic backing. As with paper chromatography, a solvent system is chosen that will best serve to isolate the photoproduct of interest. The technique holds a great deal of promise because it is simple and rapid to use, with times of only 3 to 4 hr required for good separation. In general the R_f values of the photoproducts for a given solvent are approximately the same as on paper (Greenstock and Johns, 1968).

c. Column Chromatography. Ion-exchange chromatography with Dowex columns has also been used to separate the photoproducts obtained from hydrolyzates of irradiated DNA (Wacker, 1963). These columns have several advantages over the two preceding methods of separating photoproducts. First, because there is no limit to the size of a column, one can use them to prepare large amounts of photoproduct. Second, it may be possible to separate photoproducts that tend to run together if paper or thin-layer chromatography is used. For example, many of the solvents commonly used in paper chromatography do not separate dihydrothymine from thymine, but on Dowex columns the two molecules are very easy to separate (Yamane *et al.*, 1967). Dowex column chromatography has also proved useful in separating PO-T, the deamination product of the cytosine–thymine adduct (see Section IV) from the *cis–syn* thymine dimer ($\widehat{TT}_1$) (Varghese and Day, 1970).

C. Electrophoresis

In order to measure photoproducts of cytosine that are very unstable at room temperature, Johns and co-workers (1965) have developed a separation technique using high-speed electrophoresis at low temperatures. With this procedure they have managed to separate photoproducts on paper in times of up to 15–40 min at temperatures never exceeding 10°C. Audioradiography on the intact electrophoretograms was used to detect photoproducts.

IV. DNA Photoproducts and Their Properties

A. Introduction

In general, the pyrimidines found in DNA, thymine, and cytosine

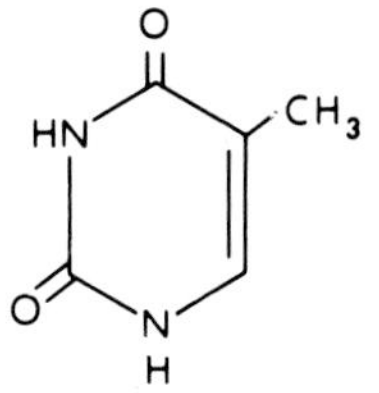

THYMINE 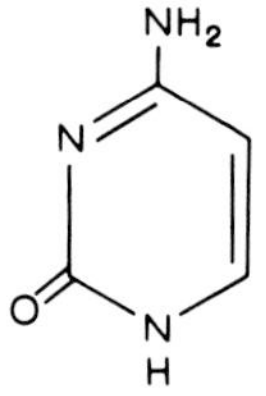 CYTOSINE

undergo photochemical reactions much more readily than the purines adenine and guanine. Hence, this discussion will focus on the various pyrimidine photoproducts that have been isolated from irradiated DNA. The photochemistry of pyrimidines is mainly characterized by addition reactions across the 5–6 double bond. Excited triplet state studies of thymine demonstrate that the excitation energy is highly localized at the 5–6 double bond (Shulman and Rahn, 1966). Theoretical calculations of the first excited state of thymine by Snyder *et al.* (1970) show that the excited state is highly localized at the C-5 to C-6 double bond. This high degree of localization of the π-π^* excited state at the 5–6 double bond has also been calculated for other pyrimidines, and Pullman (1965) has suggested that this property of pyrimidines may in part account for their photoreactivity.

The DNA photoproducts that have received the most attention in recent years are the pyrimidine dimers, in particular the thymine–thymine dimer. Dimers are readily formed in DNA and have been shown to play an important role in the lethal and mutagenic effects of UV on biological systems. In addition, much interest has been shown in the enzymatic repair systems specific for pyrimidine dimers. Other photoproducts formed in DNA are, under most conditions, less important than dimers. However, as we shall see, the relative amounts of photoproducts formed in DNA depend to a large extent upon the state of the DNA during irradiation, and under some conditions these other photoproducts may be biologically more important than dimers.

B. Cyclobutane Dimers

Thymine-containing dimers were first isolated from irradiated frozen solutions of thymine and characterized by Beukers and Berends (1960). As pointed out by Fraenkel and Wulff (1961), there are four geometrical isomers of the thymine–thymine dimer (Fig. 5). Weinblum and Johns (1966) isolated all four isomeric forms and characterized them on the

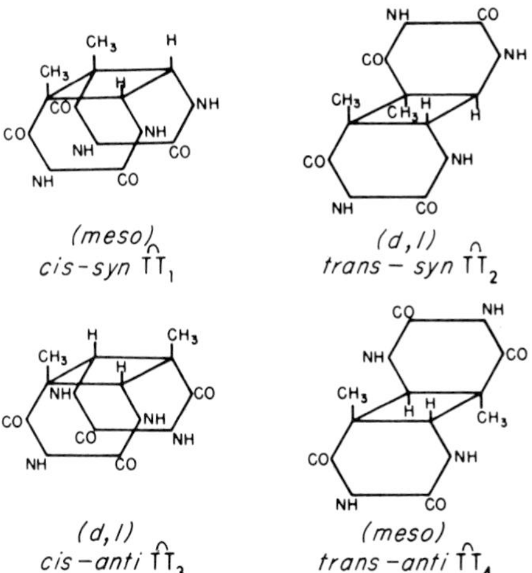

Fig. 5. The four isomers of the thymine–thymine dimer. (From Fraenkel and Wulff, 1961.)

basis of their chromatographic behavior, infrared spectra, and acid stability. The dimers $\widehat{TT}_3$ and $\widehat{TT}_4$ are not stable to the hydrolysis conditions normally used to hydrolyze irradiated DNA.

The structure of the dimer isolated from frozen solutions of thymine, as determined by X-ray crystallography, is that of the *cis–syn* dimer $\widehat{TT}_1$ (Wei and Einstein, 1968). This structure is compatible with the arrangement of the thymine molecules in a crystal of thymine monohydrate. Such crystals appear when a thymine solution is frozen (Davis and Tinoco, 1966). The dimer isolated from native DNA is also the *cis–syn* isomer, as determined by chemical analysis (Blackburn and Davies, 1966) and infrared analysis (Weinblum, 1967). According to Nagata *et al.* (1965), the *cis–syn* dimer is the dimer most likely to be formed in native DNA, because it requires the least amount of distortion of the helical structure for its formation.

Cytosine–cytosine ($\widehat{CC}$) and cytosine–thymine ($\widehat{CT}$) cyclobutane dimers are also formed upon irradiation of native DNA (R B. Setlow and Carrier, 1966). As mentioned previously, acid hydrolysis deaminates cytosine residues when the C-5 to C-6 double bond is saturated; thus the initially formed $\widehat{CC}$ and $\widehat{CT}$ dimers are isolated as $\widehat{UU}$ and $\widehat{UT}$ dimers. The relative amounts of pyrimidine dimers plus other photoproducts formed in *E. coli* DNA are given in Table II. The crystal structures of the $\widehat{UU}$ and $\widehat{UT}$ dimers isolated from DNA have yet to be determined, but the structure of a $\widehat{UU}$ dimer isolated from frozen irradiated solutions

TABLE II

Relative Yield of Photoproducts in Native and Denatured *E. coli* DNA and Their Reversibility (Indicated by + or —) with UV or Heat

	Relative yield in *E. coli* DNA			Reversibility	
Product	Native	Denatured	Ref.[a]	UV	Heat
$\widehat{\mathrm{TT}}_1$	1	1.3	1	+	−
$\widehat{\mathrm{TT}}_2$	<0.02	0.2	2	+	−
Spore photoproduct	<0.01	<0.01	3	−	−
PO-T	0.12		4	+	−
Cytosine hydrate	0.01	0.1	5	−	+
$\widehat{\mathrm{UT}}$	0.3	0.3	1	+	−
$\widehat{\mathrm{UU}}$	0.2		1	+	−
Cross-links	~0.001		6	−	−
Chain break	~0.001		6	−	−

[a] References: 1. R. B. Setlow and Carrier (1966); 2. Ben-Hur and Ben-Ishai (1968); 3. Rahn and Hosszu (1968); 4. Patrick, 1970; 5. R. B. Setlow and Carrier (1970); 6. R. B. Setlow (1968b).

of uracil has been solved and found to be the *cis–syn* isomer (Adman *et al.*, 1968). Furthermore, Weinblum (1967) showed, on the basis of infrared analysis, that the $\widehat{\mathrm{UT}}$ isomer isolated from DNA is the same as that produced upon irradiation of a frozen mixture of uracil and thymine. Since the $\widehat{\mathrm{UT}}$ dimer is probably the *cis–syn* isomer by analogy with the ice dimer $\widehat{\mathrm{TT}}_1$, Weinblum (1967) concludes that the $\widehat{\mathrm{CT}}$ dimer formed in DNA is most likely the *cis–syn* isomer.

The formation of a cyclobutane dimer results in the loss of the C-5 to C-6 double bond and the disappearance of the normal pyrimidine absorbance in the UV. The cyclobutane dimer absorbs to much shorter wavelengths than nucleic acids (Fig. 6). This absorption is due mainly to the diamide portion of the molecule, —C(O)—NH—C(O)—NH—.

One of the most important properties of cyclobutane dimers is their photoreversibility. Absorption of radiation by any of the pyrimidine dimers destroys the cyclobutane ring and restores the C-5 to C-6 double bonds of the two pyrimidine residues. The formation and reversal of the thymine dimer can be written as

$$\text{—TT—} \underset{h\nu}{\overset{h\nu}{\rightleftarrows}} \text{—}\widehat{\mathrm{TT}}\text{—}$$

The quantum yields for dimer formation and reversal at varying wavelengths of irradiation are given in Table III. As indicated, the yields vary

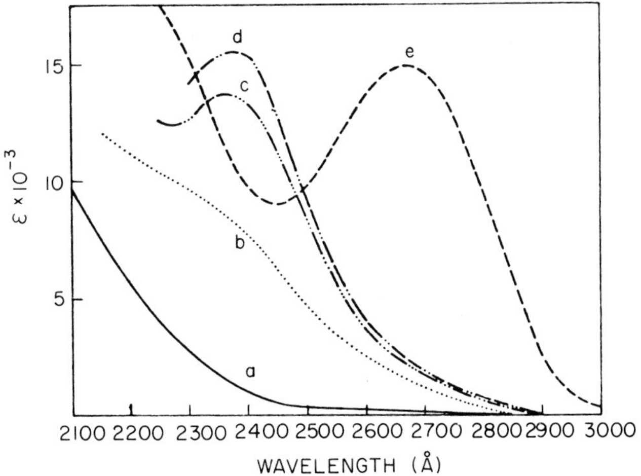

Fig. 6. Absorption spectrum of the photo-induced p$\widehat{T\text{-}T}$ dimer at neutral (*a*) and various alkaline pH values taken from Sztumpf and Shugar (1962); (*b*) pH 12; (*c*) pH 13; (*d*) pH 14. Also shown is the spectrum following photoreversal (*e*), which corresponds accurately to the alkaline absorption spectrum of p$\widehat{T\text{-}T}$.

slightly for irradiation between 235 and 275 nm. Also shown are the extinction coefficients for a thymine pair in DNA and for the isolated $\widehat{TT}_1$ dimer. Obviously, the absorbance of thymine is much greater at long wavelengths than that of the dimer, whereas at short wavelengths the dimer absorbs to the same extent as thymine. The photosteady values of $\widehat{TT}_1$ at various wavelengths of irradiation reflect this variation in molar extinction between thymine pairs and the thymine dimer. As shown in Table III, the photosteady values of $\widehat{TT}_1$ in *E. coli* DNA range from 1.7% for irradiation at 235 nm to 20% for irradiation at 280 nm. Clearly, the low yield at 235 nm reflects the increase in the absorbance of the

TABLE III

Wavelength Dependence of Quantum Yields (Φ) and Extinction Coefficients (E)

Excitation wavelength (nm)	Φ Dimer[a] formation	Φ Dimer[b] reversal	E[a]	E[b] ($\widehat{TT}_1$)
235	0.036	0.8	2,800	1,500
254	0.033	0.65	8,200	300
275	0.023	0.6	10,200	18

[a] For adjacent thymines in *E. coli* DNA. (From Wulff, 1963.)
[b] For isolated thymine dimers. (From Herbert *et al.*, 1969.)

dimer at short wavelengths relative to thymine itself. This larger absorbance at short wavelengths leads to a much higher rate of dimer reversal and lower photosteady yields of dimer.

The ability of dimers to undergo short wavelength reversal is useful in characterizing the presence of a cyclobutane ring. As an illustration, irradiation at 235 nm of *E. coli* DNA containing a thymine dimer yield of greater than 1.7% leads to a reduction in dimer yield until the photosteady concentration of 1.7% is reached. The dimers are completely lost when hydrolyzates of irradiated DNA are irradiated in solution. In this instance the two thymine residues making up the dimer are not connected by means of a phosphodiester linkage. Hence, upon reversal of the isolated dimer, the cyclobutane ring is destroyed and since there is nothing holding the two thymine molecules together they move apart. The quantum yield of dimerization for thymine in solution is very low, so very few of the split dimers reform. This means of "direct" reversal is distinct from that of photoenzymatic reversal obtained with light of wavelengths greater than 300 nm in the presence of the photoreactivating enzyme (Section V).

The photosteady yields of $\widehat{TT}_1$ given in Table IV for *E. coli* DNA are considerably less than for poly(dT). This difference reflects the fact that in DNA not all the thymines have a thymine neighbor available for dimerization as does poly(dT). On the average, in *E. coli* DNA 50% of the thymines have a thymine neighbor on the same strand, but this 50% includes sequences of the form —TTT—, which can only form one dimer. It is estimated, therefore, that about 44% of the thymine in *E. coli* DNA

TABLE IV

Steady-State Thymine Dimer ($\widehat{TT}_1$) Yields in Two Different DNA's and in Poly(dT) at Various Wavelengths of Radiation

Wavelength (nm)	Percentage thymine as $\widehat{TT}_1 + \widehat{TT}_2$		
	E. coli DNA[a]	*H. influenzae* DNA[b]	Poly (dT)[c]
235	1.7	—	14
254	6.5	13	39
280	20	32	66
Acetophenone sensitization ($\lambda > 300$)	37	44	80

[a] $(A + T)/(G + C) = 0.93$.
[b] $(A + T)/(G + C) = 1.63$.
[c] Except for acetophenone sensitization, from Deering and Setlow (1963).

can actually be converted to $\widehat{TT}$ dimers. For a given DNA the maximum percentage of thymine converted to $\widehat{TT}_1$ also reflects the $(A + T)/(G + C)$ ratio of that DNA. As shown in Table IV, the photosteady yield of $\widehat{TT}_1$ for *Haemophilus influenzae* DNA $[(A + T)/(G + C) = 1.63]$ is greater than that of *E. coli* DNA $[(A + T)/(G + C) = 0.93]$ for a variety of wavelengths. This difference represents the higher frequency of thymine-thymine pairs in the *H. influenzae* DNA.

Although it is not possible to dimerize two pyrimidine residues separated by a purine residue, as in poly(dA-dT), for example, preferential removal of the adenine residues by mild hydrolysis gives a polymer of the form —T—s—T—s—T— (s = deoxyribose). Now the thymine residues are adjacent and undergo dimerization, but at a rate five times less than that of poly(dT) (Rahn and Landry, 1971). This decrease in rate presumably reflects a greater time-averaged distance between the thymine residues in —T—s—T— relative to —T—T—.

Another thymine dimer isomer, the *trans–syn* isomer $\widehat{TT}_2$, has been isolated from irradiated denatured DNA by Ben-Hur and Ben-Ishai (1968). It has also been isolated from irradiated TpT by Johns *et al.* (1964), who showed that the quantum yield of $\widehat{TT}_2$ formation is about five times less than that of $\widehat{TT}_1$. Thus, the ratio of $\widehat{TT}_1$ to $\widehat{TT}_2$ is about 5:1. Interestingly, similar ratios have been found in other systems, as indicated in Table V. These systems are characterized by a high degree of disorganization or base unstacking. The structure of $\widehat{TT}_2$ requires a rotation of one base with respect to another. Apparently this ability to rotate is restricted in the double helical conformation. As indicated in

TABLE V

Ratio of *trans–syn* to *cis–syn* Thymine Dimers ($\widehat{TT}_2/\widehat{TT}_1$) in a Variety of Thymine-Containing Systems[a]

	$\widehat{TT}_2/\widehat{TT}_1$
TpT[c]	0.20
Poly(d-dT)[b]	0.19
Apurinic acid	0.18
Poly(dT)	0.15
Denatured DNA	0.14
Poly(dA-dT)	0.045
Native DNA	<0.02

[a] Rahn and Landry (1971).
[b] Depurinated poly(dA-dT).
[c] Johns *et al.* (1964).

Table V, the incorporation of random coil poly(dT) into the ordered poly(dA)·poly(dT) structure leads to a large reduction in the rate of $\widehat{TT}_2$ formation.

Several groups of workers have found that low pH favors thymine dimerization in DNA irradiated at 313 nm. Haug and Sauerbier (1965) observed that the rate of dimerization of thymine in phage was 2.5 times greater at pH 3.5 than at pH 7 in the presence of oxygen, and was 5.2 times greater in the presence of nitrogen. R. B. Setlow and Carrier (1966) also found that for a constant dose of 313 nm radiation the percentage of thymine dimer is isolated DNA increased from 0.22 to 3.5% when the pH was lowered from 8.0 to 3.4. Although interesting, these results have not been satisfactorily explained.

A proton is lost from the N-3 position of thymine with a pK_a of 9.8. Studies on poly(dT) have shown that the rate of dimerization is drastically reduced upon loss of this proton (Rahn, 1970). The reduction may be due to electrostatic repulsion between the negatively charged bases. Loss of the N-3 proton also leads to a large shift in the absorbance of the dimer (Fig. 6). Consequently, the amount of light absorbed by the dimer relative to the monomer increases upon ionization, and the rate of dimer reversal is thereby enhanced, provided the quantum yield does not decrease upon deprotonation. Therefore, loss of the N-3 proton of thymine in poly(dT) or DNA shifts the photosteady equilibrium in favor of the monomer by both decreasing the rate of dimer formation and increasing the absorbance of the dimer. As shown in Fig. 6, irradiation of the $\widehat{pTpT}$ dimer at pH 13 with 254 nm radiation completely reverses the dimer. Similar results have been obtained with poly(dT) (Rahn, 1970).

The formation of thymine dimers in DNA has been studied as a function of the temperature of the sample during irradiation over a wide range of temperatures. Usually the rates of photo-induced chemical processes are not influenced by temperature, i.e., the energy of activation is zero, provided the overall reaction is limited by the rate of the purely photochemical step and not by factors involving molecular motion, such as diffusion or rotation (Giese, 1962). The reason is that the energy of an absorbed photon is so much greater than the Boltzman factor KT in the normal experimental temperature range that increasing the energy of a molecule by increasing the temperature has little effect on the chemical reactivity of the excited molecule. For example, the quantum yield of thymine dimerization in frozen solutions of thymine is independent of temperature (Eisinger and Shulman, 1967; Füchtbauer and Mazur, 1966). This result implies that very little motion is required for the two pyrimidines to form a dimer and that the free space required for motion represents a small fraction of the total space occupied by the two interacting molecules. In DNA, on the other hand, the thymine residues are attached to a sugar phosphate backbone and consequently are further apart than in the thymine crystal. Hence, the degree of interaction between two adjacent thymines in the polymer is considerably less than in the crystal. Consequently, the quantum yield of thymine dimerization in DNA is considerably less than in frozen solutions of thymine. It is to be expected, then, that structural and environmental factors which influence the ability of two thymines to interact will influence the rate of dimerization and that such factors are affected by the temperature.

An example of the influence of temperature on the rate of dimerization is shown in Fig. 7, where the yield of thymine dimers is given for DNA irradiated at temperatures ranging from 25° to 90°C. At 80°C the DNA melts cooperatively, going from the double-stranded to the single-stranded form. The rate of dimerization also undergoes a sharp decrease at the melting temperature. This decrease in dimerization at high temperatures means that forming a dimer in the unstacked, single-stranded form of DNA (above T_m) is more difficult than in the stacked, double-stranded form (below T_m). Single-stranded DNA is highly stacked at low temperatures, however, and the rate of dimerization at 25°C in heat-denatured DNA is 30% greater than in native DNA. However, an increase in the temperature decreases the base stacking in a noncooperative manner and the rate of dimerization correspondingly decreases in a linear fashion (Fig. 7). Stacking appears to have a much greater influence on photo-dimerization than strandedness.

The photochemistry of DNA has also been studied at temperatures

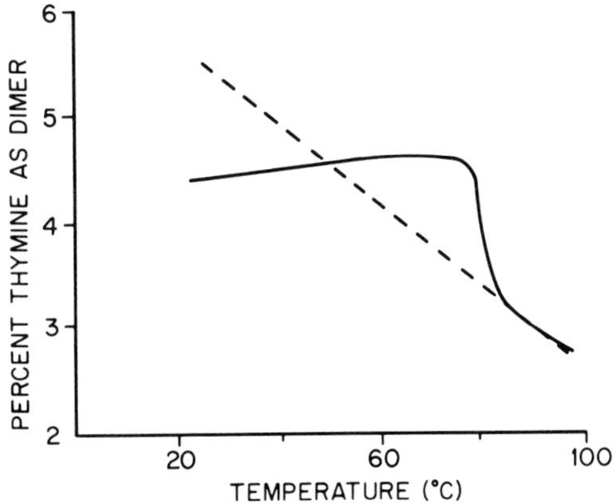

Fig. 7. Variation in thymine dimer yield as a function of temperature (25°–100°C) for a fixed dose of irradiation. Native DNA (—); denatured DNA (----). (Adapted from Hosszu and Rahn, 1967.)

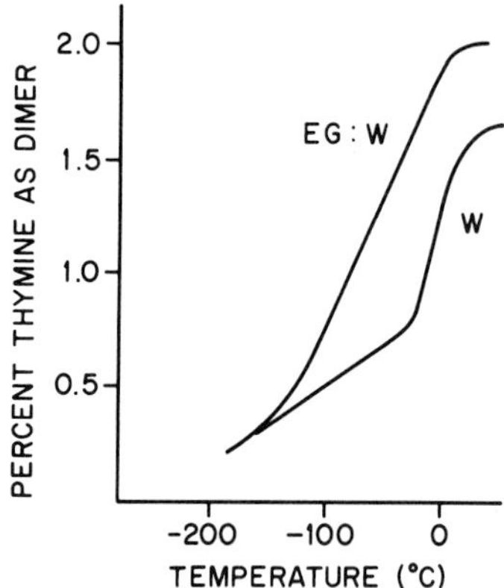

Fig. 8. Variation in thymine dimer yield as a function of temperature (—196° to 25°C) for a fixed dose (1×10^4 ergs·mm^2) of irradiation at 280 nm. DNA was either in water or in a 1:1 mixture of ethylene glycol and water. (Adapted from Rahn and Hosszu, 1968.)

ranging from 25° to —196°C. As shown in Fig. 8, the yield of thymine dimers formed in DNA decreases with decreasing temperatures of irradiation. In water, the decrease in yield is sharpest just below the freezing temperature. On the other hand, the yield in 50% ethylene glycol decreases linearly with decreasing temperature down to —130°C, the solidification temperature of this solvent (Luyet and Rasmussen, 1968). The difference in behavior of the rate of dimerization in these two solvents suggests that dimerization may be, in part, a viscosity-controlled reaction (Gegiou *et al.*, 1968). In any case, the decrease in dimer yield at very low temperatures is attributed to the solidification of the solvent, which makes it more difficult for two adjacent thymines to attain the proper orientation for dimerization. In a rigid matrix, this freedom of motion is diminished and extensive irradiation (280 nm) at —196°C does not lead to more than 1% dimer. However, reirradiation of a sample of DNA that has been thawed and refrozen leads to further dimerization, suggesting that freezing of DNA locks into place a small percentage of the thymine in orientations suitable for dimerization. Repeated thawing and refreezing produces a new population of adjacent thymines that can undergo dimerization upon reirradiation.

It is of interest that the initial rate of dimerization in poly(dT) is from five to ten times faster at —196°C than at 25°C, but only 30% of the thymines can dimerize at —196°C (provided there is no intermittent thawing). Therefore, the rate of dimerization at —196°C for this 30% is roughly 20 to 30 times greater than the rate at 25°C, and the quantum yield of dimerization approaches unity while the yield for the remaining 70% is approximately zero. Since the quantum yield for dimerization of thymine in ice is close to unity (Eisinger and Shulman, 1967) it appears that a portion of the thymines in poly(dT) at —196°C are in an environment that approximates that found in frozen solutions of thymine.

C. SPORE PHOTOPRODUCT (5-THYMINYL 5,6-DIHYDROTHYMINE)

It has been known for some time that bacterial spores are very resistant to heat as well as to the action of ultraviolet radiation (Becquerel, 1910). Their resistance to UV may in part be due to the fact that irradiation of spores does not form thymine dimers but forms instead a rather unique photoproduct, 5-thyminyl 5,6-dihydrothymine, which like the pyrimidine dimer is made up of two adjacent thymines (Varghese, 1970).

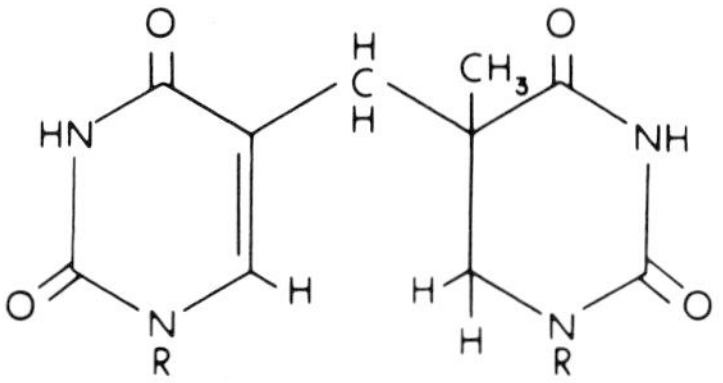

Donnellan and Setlow (1965) first isolated the spore photoproduct from UV-irradiated bacterial spores of *Bacillus megaterium*. They found that its maximum yield was about 40% of the total thymine, a percentage approaching the maximum number of thymine–thymine pairs.

Since the spore photoproduct has one intact chromaphore, it maintains an absorbance equal to that of a single thymine residue. Unlike the thymine dimer, the spore photoproduct has no cyclobutane ring and is not reversed when irradiated in solution. The mechanism leading to the formation of the spore photoproduct has been proposed to involve a free radical intermediate of thymine formed by the removal of a hydrogen atom from the methyl group (Varghese, 1970). This radical then attacks a neighboring thymine and adds across the C-5 to C-6 double bond.

The spore photoproduct, which is readily formed at 25°C in spores, is not formed in either vegetative cells or in isolated DNA under the same conditions of irradiation. At low temperatures, however, it is readily formed in both isolated DNA (Rahn and Hosszu, 1968) and bacterial cells (Smith and Yoshikawa, 1966). The yield of the spore photoproduct in DNA as a function of temperature reaches a maximum at about —100°C (Fig. 9). The yield decreases with decreasing temperature until by —196°C only a small percentage of thymine is converted to spore photoproduct.

The formation of the spore photoproduct in DNA is also favored by conditions of low relative humidity. Films of DNA were equilibrated at

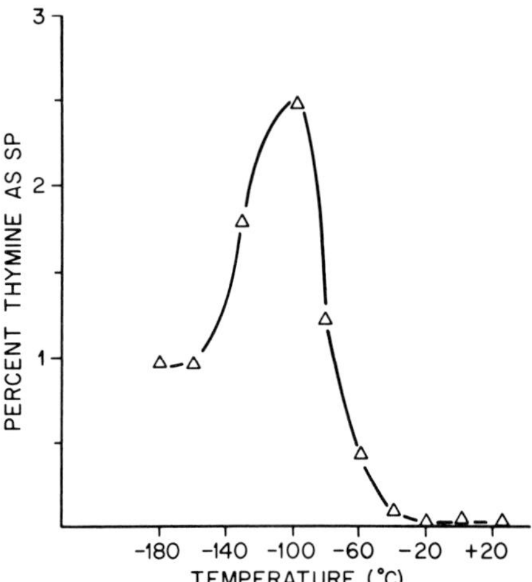

Fig. 9. Variation in the yield of the spore photoproduct (SP) in DNA as a function of the temperature of irradiation. (Adapted from Rahn *et al.*, 1969.)

various relative humidities, irradiated, and the yields of photoproducts measured (Rahn and Hosszu, 1969). Figure 10 shows a very sharp increase in the yield of spore photoproduct for a constant dose when the relative humidity is decreased below approximately 65%. There is also a corresponding decrease in the yield of thymine dimer below this relative humidity. Optical studies on films of DNA (Falk *et al.*, 1963) have shown that at approximately 65% relative humidity DNA undergoes a coopera-

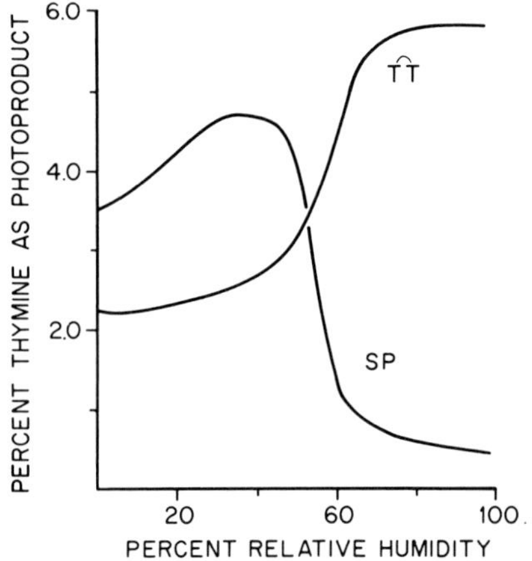

Fig. 10. Photoproduct yields in DNA films equilibrated at various relative humidities and irradiated (254 nm) with 60,000 ergs·mm^{-2}. The films were kept at 0% relative humidity before equilibration. (Adapted from Rahn and Hosszu, 1969.)

tive transition from an ordered form to a disordered form. Presumably, the spore photoproduct is favored by the DNA structure that prevails in a partially dried and disordered state. Such a dry state may exist in spores as well as in DNA at low temperatures; some of the water bound to the DNA may be frozen out as ice crystals. The recently proposed mechanism for spore photoproduct formation involving a free radical intermediate (Varghese, 1970) would be favored by the absence of water. Hence, the "dry" state of DNA may be an important reason for the formation of spore photoproducts in spores, frozen DNA, and dry films of DNA.

D. Pyrimidine Adducts

Another class of photoproducts formed between two adjacent pyrimidines are the pyrimidine adducts. Varghese and Wang (1967) first reported the presence of a thymine-containing photoproduct in acid hydrolyzates of UV-irradiated DNA that was later shown (Wang and Varghese, 1967), on the basis of infrared and NMR data, to have the structure given by IV in Fig. 11. Wang and Varghese (1967) proposed

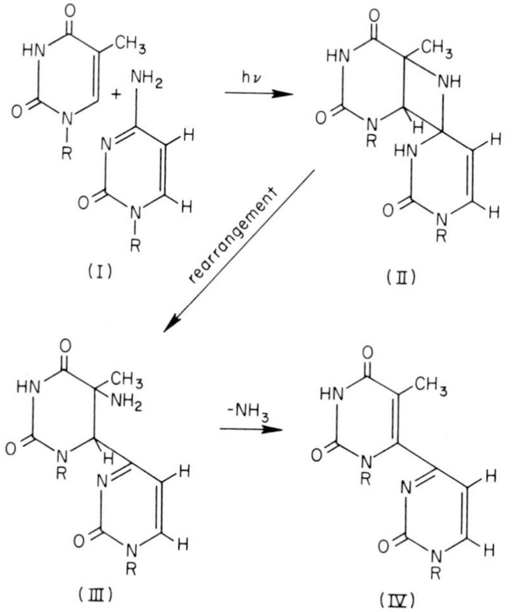

Fig. 11. Proposed mechanism (Wang and Varghese, 1967) for cytosine–thymine adduct (III) formation in DNA. The product isolated from hydrolyzed DNA, PO-T, is given by (IV).

that irradiation of DNA leads initially to the formation of an azetidine derivative (II) of thymine and cytosine. This derivative is presumably unstable and rearranges to form product III, which then undergoes deamination upon hydrolysis to give IV, the acid-stable product, 6-4′-[pyrimidin-2′-one]-thymine (PO-T). This photoproduct chromatographs in a large variety of solvents with the same R_f as the thymine dimer $\widehat{TT}_1$. Since $\widehat{TT}_1$ is normally present in considerably large quantities than PO-T, the presence of PO-T was never determined until Varghese and Wang (1967) found a suitable solvent for its chromatographic separation (see Table I).

The corresponding thymine–thymine adduct

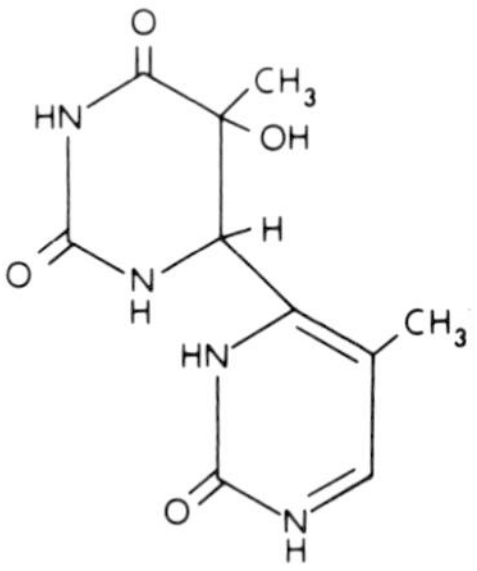

has been obtained from irradiated frozen solutions of thymine (Varghese and Wang, 1968) but not from DNA, and its crystal structure determined (Karle *et al.*, 1969).

The adducts, because of the pyrimidin-2′-one moiety, have an absorbance maximum at 315 nm (Fig. 12). An increase in the absorbance at 315 nm upon irradiation of DNA has been attributed to the formation of the cytosine–thymine adduct (Varghese and Wang, 1967). When PO-T is irradiated at 315 nm, it is converted into an unknown product with an absorbance resembling thymine (Varghese and Wang, 1967).

According to Patrick (1970), the amount of PO-T formed in DNA depends upon the $(A + T)/(G + C)$ ratio of the DNA. The amount of PO-T formed relative to thymine dimer in DNA's from several different sources is shown in Table VI. It is clear that DNA with a high cytosine content has a much higher yield of the cytosine–thymine adduct.

E. Cytosine Hydrate

The preceding sections have dealt solely with photochemical interactions between adjacent pyrimidines. In this section we consider the photo-

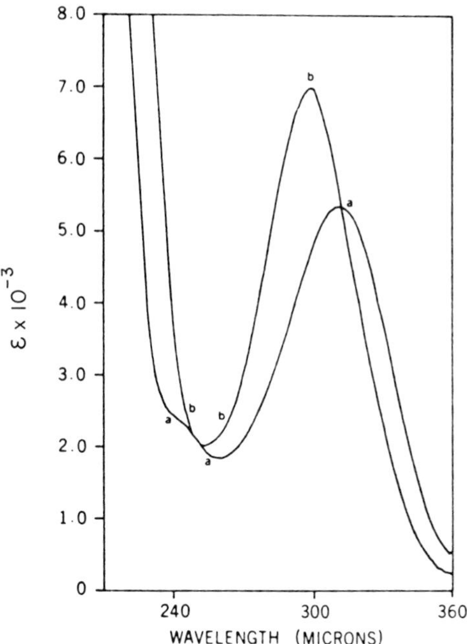

Fig. 12. Ultraviolet absorption spectra of isolated PO-T in (a) water or 0.1 *N* HCl and (b) 0.1 *N* NaOH. (From Wang and Varghese, 1967.)

chemical interaction between cytosine and water, i.e., the formation of the cytosine hydrate. The photochemical addition of water across the 5–6 double bond of uracil was first observed by Sinsheimer and Hastings (1949) and was one of the very earliest indications that nucleic acids

TABLE VI
Ratio of PO-T Adducts to *cis–syn* Thymine Dimers in Three DNA's Varying in (A + T)/(G + C) Ratio[a]

	(A + T)/(G + C)	$PO\text{-}T/\widehat{TT}_1$
E. coli B/r	0.93	0.12
H. influenzae	1.63	0.08
Micrococcus radiodurans	0.51	0.32

[a] Patrick, private communication (1970).

form well-defined photoproducts. A similar water-addition product is formed by cytosine but not by thymine. The structure of the cytosine hydrate obtained from irradiated solutions of cytidine (Miller and Cerutti, 1968) is

This photoproduct is unstable at room temperature and reverses back to the monomer quite readily. The cytosine hydrate shows an absorbance peak at 240 nm that is characteristic of dihydrocytosine derivatives (Fig. 13). In addition, conversion of cytosine to the hydrate results in the loss of the characteristic absorbance peak (270 nm). The formation and reversal of the hydrate, therefore, can in some cases be easily followed spectrophotometrically. Further, on the basis that triplet sensitization of uracil in solution does not lead to any hydrate formation (Lamola and Mittal, 1966), one concludes that photohydration of cytosine in DNA proceeds through the singlet state and not the triplet state.

Grossman and Rogers (1968) first pointed out the usefulness of hydrogen exchange as a quantitative method for measuring the formation of the cytosine photohydrate in DNA. [Previous workers (Shapiro and Klein, 1967) had shown that the C-5 hydrogen of dihydrocytosine and

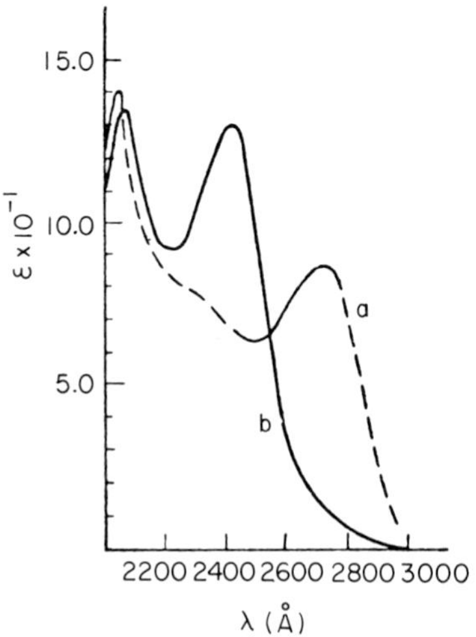

Fig. 13. Absorption spectra of (a) cytosine monophosphate and (b) photohydrate of cytosine monophosphate. (From Wierzchowski and Shugar, 1961.)

related compounds exhibited exchange reactions with water.] By preparing DNA labeled with 5-^{3}H-cytosine, Grossman and Rogers were able to measure this exchange. They absorbed the irradiated DNA on activated charcoal, filtered the suspension, and counted the activity of the filtrate. They claimed that for every hydrate formed there was a quantitative release of label. This lability is based on the following proposed tautomer of the cytosine hydrate involving a double bond at the 4–5 position (Grossman and Rogers, 1968):

NH$_2$ / N / H / H(T) / H / OH / O / N / R ⇌ NH / HN / H / H(T) / H / OH / O / N / R ⇌ NH$_2$(T) / (T)HN / H(T) / H / OH / O / N / R

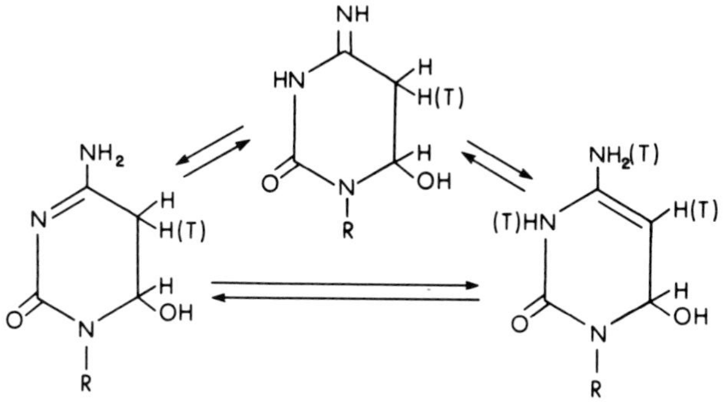

It is clear from the work of Grossman and Rogers (1968) and R. B. Setlow and Carrier (1970) that photohydration of cytosine in DNA proceeds ten times faster when the DNA is denatured. This result is consistent with the observation of Pearson and Johns (1966) on the rate of photohydration of poly U relative to poly(A)·poly(U). They found that the incorporation of poly U into the double-stranded polymer reduced the rate of hydration by a factor of 10. Presumably, this reduction is due to the reduced accessability of uracil to the solvent in a double-stranded structure as compared to a single-stranded structure. At high levels of irradiation, dimer formation causes local denaturation of native DNA and the rate of cytosine hydration increases, giving a sigmoid dose–response curve (R. B. Setlow and Carrier, 1970). That is, the cytosines located in the denatured regions are much more susceptible to photohydration. R. B. Setlow and Carrier (1970) have estimated that the rate of hydrate formation in native DNA is 30 times less than the rate of dimer formation.

Recently, DeBoer and Johns (1970) measured the hydrogen exchange from hydrates of cytosine derivatives and concluded that tritium was released from the 5-position mainly by the exchange mechanism, with only a slight loss of tritium following dehydration of the hydrate. They found, in contrast to the work of Grossman and Rogers (1968), that the total release of tritium was not quantitative and concluded that the loss of tritium is not a good assay for hydrates. The reductive assay method (Section V) shows great promise for determining hydrate formation quantitatively.

F. Dihydrothymine

Recently, Yamane *et al.* (1967) isolated from extensively irradiated DNA a photoproduct they identified as dihydrothymine:

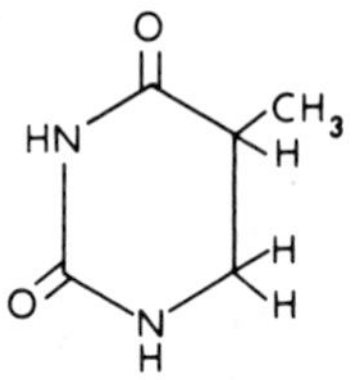

O
HN CH$_3$ H
H H
O N H

They chromatographically isolated this product on a Dowex column because paper chromatography in a variety of solvents (Table I) failed to separate thymine from dihydrothymine. The doses of radiation needed to produce observable amounts of this photoproduct are considerably greater than those required for a comparable amount of thymine dimer. Yamane and co-workers suggested that the thymine free radical may be a precursor for the formation of dihydrothymine, but it remains to be seen whether there is any correlation between the yield of free radical and the yield of dihydrothymine. Such a correlation is difficult to make because free radicals are observed only when DNA is irradiated in a dry or frozen state, i.e., under conditions that trap the radicals.

G. Cross-Links

1. *DNA Cross-Links*

DNA that has been irradiated with UV and then subjected to acid or thermal denaturation does not undergo irreversible strand separation to the same extent as unirradiated DNA according to Marmur and Grossman (1961). In their experiments UV-irradiated DNA, after thermal denaturation, still maintained double-stranded regions as judged by the fraction of DNA that banded in a CsCl density gradient as native DNA. This failure of the strands to undergo complete separation under normal denaturation conditions was attributed to the formation of cross-links between the strands.

The exact nature of the link is not known, but an indication that it involves thymine is the observation that DNA's with high A + T contents form cross-links at a faster rate than DNA's with lower A + T contents (Marmur *et al.*, 1961). If any one thymine dimer is to be favored for forming a cross-link, it is probably the *trans–anti* dimer, because it more closely resembles the relative orientation of two thymines located on opposite strands. This dimer is acid labile and cannot be isolated by acid hydrolysis. Unpublished results obtained by Drake and Setlow (1962) have indicated that the UV-induced cross-link is not short-wavelength reversible, i.e., is not a cyclobutane dimer.

Glišin and Doty (1967) have measured cross-linking in DNA as a function of both temperature and pH. They found that when DNA is approximately 20% denatured, either at low pH (approx. 3.2) or at temperatures just below the T_m, the rate of cross-linking increases by a factor of 3–4 (Fig. 14). Presumably, the increase in the flexibility of the strands in the locally denatured regions favors the formation of the cross-link. A high salt concentration (0.2 *M* NaCl) was necessary to achieve this enhancement of cross-linking in the partially denatured DNA. At tenfold lower salt concentrations there was no such enhancement. Presumably the high salt concentration provides electrostatic shielding of the negatively charged phosphates and prevents the strands from moving completely apart in the locally denatured regions. Under conditions that favor complete strand separation the rate of formation of cross-linking rapidly falls to zero. Irradiation of DNA at low temperatures in frozen media inhibited the formation of cross-links such that at —196°C no UV-induced cross-links were detected (Rahn *et al.*, 1969).

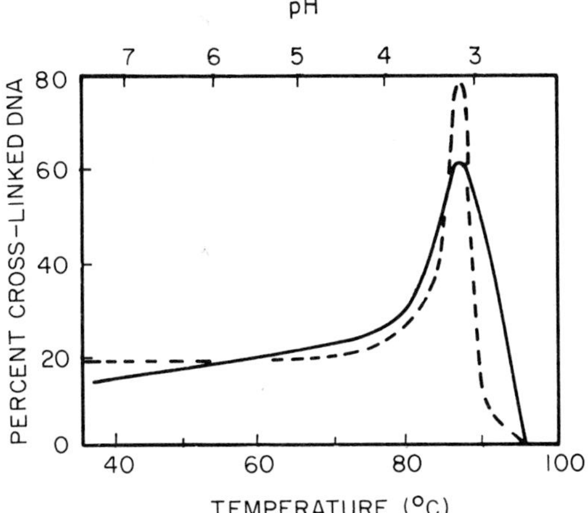

Fig. 14. UV-induced cross-link formation in DNA as a function of either temperature (—) or pH (- - - -). (Adapted from Glišin and Doty, 1967.)

TABLE VII
UV-Induced Cross-Linking, Measured by Relative Rates of Gelation, in DNA Irradiated over a Wide Range of Relative Humidities[a]

	Relative humidity (%)				
	33	40	50	75	97.5
Relative rates of gelation	22.1	5.7	3.7	—	1

[a] From K. Kaplan, 1955.

Another kind of DNA cross-linking is an intermolecular type between two different DNA molecules. Such cross-linking leads to gel formation and occurs when dry films of DNA are irradiated (R. B. Setlow and Doyle, 1953). The rate of gel formation is independent of temperature between 25° and —183°C. Gel formation is favored by low relative humidity since no gel is formed following irradiation of DNA in solution. K. Kaplan (1955) measured the rate of gelation of DNA films as a function of the relative humidity (Table VII) and found that gel formation was greatly enhanced when the relative humidity was below ~50%. There is good evidence that only the pyrimidine residues are involved in the interstrand cross-linking reaction. Baranowska and Shugar (1960) found that gel formation proceeded in films of apurinic acid as readily as in DNA. Hence, the loss of the purines has no effect on the rate of gel or fiber formation.

2. *DNA-Protein Cross-Links*

The formation of UV-induced cross-links between DNA and protein has been postulated to account for the inability to separate DNA from protein in lysates of irradiated bacterial cells (Smith, 1964). Such cross-links are also formed when DNA and bovine serum albumin are irradiated together *in vitro,* because the amount of DNA recovered from the protein decreases linearly with increasing amounts of UV-irradiation (Smith, 1967).

In order to learn more about the chemical nature of the protein–DNA cross-link, Smith and Meun (1968) irradiated ^{35}S-cysteine in the presence of various polynucleotides. They found that cysteine forms UV-induced links with polymers containing thymine and cytosine much more readily than with polymers containing adenine or guanine. Smith (1968) has speculated that the addition of cysteine to thymine in DNA might follow a reaction scheme whereby a free radical formed at thymine interacts with cysteine to form 5-*S*-cysteine-6-hydrothymine.

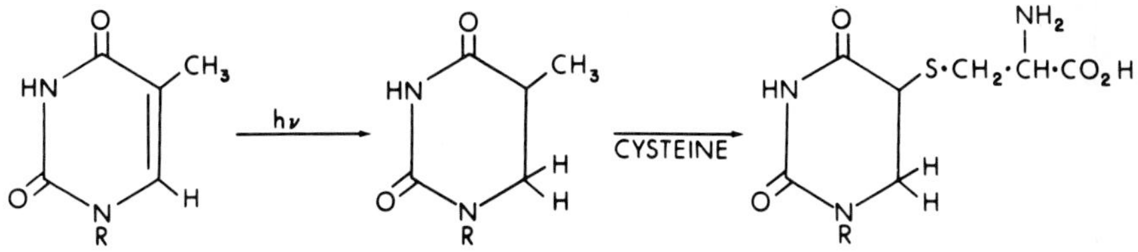

Cysteine, with its electron-deficient disulfide linkage, is known for its ability to interact with unpaired electrons and is often used as a free radical scavenger.

However, another possibility is that a free radical formed at cysteine might interact with the base. Recent studies by Jellinek and Johns (1970), in which cysteine was irradiated in the presence of uracil, have indicated that the triplet of uracil abstracts a hydrogen atom from cysteine and the resulting thiyl radical then interacts covalently with a ground state uracil molecule. A similar reaction may occur with thymine.

The existence of stable precursors for the DNA–protein cross-linking reaction is suggested by the observation (Smith, 1964) that cross-linking is obtained between DNA and bovine serum albumin even when the protein and/or the DNA are irradiated separately and then mixed. Such a precursor may be a photochemically transformed cytidine residue with a saturated 5–6 bond. Since a covalent link is formed between dihydrocytosine and glycine (Janion and Shugar, 1967) with the loss of the amino group of the dihydrocytosine, it is reasonable, as pointed out by Smith (1968), that similar reactions should occur between glycine and either the hydrate or the dimer of cytosine. Smith and O'Leary (1967) measured protein–DNA cross-links as a function of the temperature of irradiation and found that at 25° and —196°C fewer cross-links were formed in irradiated cells of *E. coli* than at —79°C. Presumably, the interaction between DNA and protein that favors cross-linking is enhanced at —79°C, possibly because the water in the cell is converted into ice crystals. The rigid environment that must prevail at —196°C could restrict the interaction between the excited state precursors and inhibit the formation of a protein–DNA cross-link.

H. Chain Breaks

Of all the lesions formed upon UV-irradiation of DNA, probably least is known about chain breaks. Alexander and Moroson (1960) found that extensive UV-irradiation of DNA caused a decrease in light scattering, which they attributed to a reduction in molecular weight caused by chain breaks. Since they looked only at native DNA, the decrease in molecular weight must have been due, for the most part, to breaks in both strands occurring within a short distance of one another. Marmur *et al.* (1961), by first denaturing the DNA and then measuring the decrease in molecular weight by sedimentation analysis, were able to show that UV produced single-strand breaks in native DNA. Interestingly, they were not able to show that irradiation of single-stranded DNA caused chain breaks.

A complicating factor in measuring chain breaks is the possibility that cross-links may be holding the strands together, so the actual number of breaks may be greater than the number apparent from changes in molecular weight. Hence, studies in which both chain breaks and cross-links are measured are desirable. Such a study was made by Bujard (1970) on DNA from bovine papilloma virus. This DNA is a double-stranded, covalently closed circular molecule that sediments under alkaline conditions at 64 S. One single-strand break in the molecule produces a single-stranded linear form (20 S) and a single-stranded circular form (23 S). Hence, the loss of the 64 S peak is a sensitive measure of strand breakage. On the other hand, a combination of one break and one cross-link gives a species consisting of a circular strand cross-linked to a linear strand and having an S of 30. The appearance of this new band migrating with 30 S is a measure of interstrand cross-links. From these studies it was determined that 1 erg$\cdot$mm^{-2} at 280 nm produced 2.3×10^{-6} single-strand breaks and 1.6×10^{-6} interstrand cross-links per 1×10^{6} daltons of double-stranded DNA.

Chain breaks are readily formed upon irradiation of DNA that contains 5-bromouracil (BrUra) in place of thymine. The formation of chain breaks in BrUra-labeled DNA will be discussed separately in Section VI,A.

I. Biological Significance of Photoproducts

We have made no attempt to classify photoproducts on the basis of their biological importance. Studies devoted to this task are complicated by the possibility of biological repair of the photodamage. However, it is generally thought (R. B. Setlow, 1968b) that a single cross-link or chain break is two to three orders of magnitude more lethal than a single pyrimidine dimer or hydrate. Also, there is evidence that pyrimidine dimers are from five to ten times more lethal than the spore photoproduct (Rahn *et al.*, 1969; Stafford and Donnellan, 1968). The biological significance of the cytosine–thymine adduct is still uncertain even though Varghese and Day (1970) have shown that it is excised from irradiated cells of *Micrococcus radiodurans* at the same rate as pyrimidine dimers. In general, since the ability to repair photodamage varies from system to system, the relative importance of a particular type of photoproduct may very well depend upon the system studied.

V. Physical and Biological Properties of Irradiated DNA

A. Physical Properties

The formation of photoproducts in DNA disrupts hydrogen bonding and decreases base stacking in the vicinity of the photoproduct. The presence of these locally denatured regions in an irradiated DNA molecule manifests itself in several ways. Marmur *et al.* (1961) observed that the melting temperature of DNA decreases roughly in proportion to the amount of UV-radiation (254 nm) absorbed by the DNA. Photoproducts other than pyrimidine dimers contribute to this lowering of T_m because reversal of the pyrimidine dimers by short-wavelength irradiation fails to produce an equivalent change in the melting temperature of the DNA (R. B. Setlow and Carrier, 1963). Furthermore, irradiation of DNA containing a steady-state concentration of dimers leads to a further decrease in the T_m, even though the concentration of dimers remains constant. On the other hand, acetophenone sensitization leads nearly exclusively to thymine dimer formation (Lamola and Yamane, 1967) and, as shown in Fig. 15, there is a linear relationship between the thymine dimer content of the DNA and the decrease in the melting temperature. Conversion of approximately 5% of the thymine to dimers lowers the melting temperature by 1°C.

Extensive irradiation of DNA leads to complete denaturation if the irradiation temperature is sufficiently close to the melting temperature of the DNA. For example, extensive irradiation of native DNA at 25°C in 90% ethylene glycol, in which the T_m of DNA is 35°C, results in complete denaturation of the DNA (Rahn and Landry, 1969). Denaturation takes place regardless of whether the DNA is irradiated in the presence of glycol or whether the glycol is added after the irradiation. So the glycol only serves to lower the T_m and does not participate in any chemical reaction with the DNA.

Another indication of the presence of denatured regions in irradiated DNA is the increase in the rate of reaction of DNA with formaldehyde after UV-irradiation (Marmur *et al.*, 1961). Formaldehyde reacts much more quickly with single-stranded DNA than with double-stranded DNA

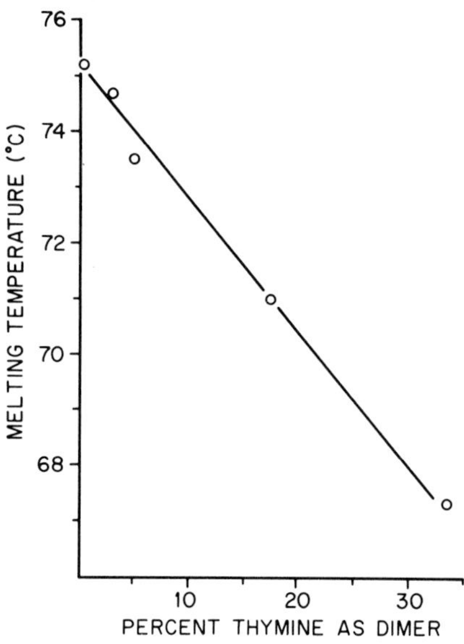

Fig. 15. Correlation in DNA between thymine dimer yield and the melting temperature following acetophenone sensitization. (From Rahn and Landry, unpublished results, 1970.)

(Grossman *et al.*, 1961). Trifonov *et al.* (1968) have developed a kinetic theory for analyzing the interaction between UV-irradiated DNA and formaldehyde.

The interaction of acridine orange with irradiated DNA has also been used to follow the UV-induced formation of locally denatured regions in DNA (Zavil'gel'skii *et al.*, 1964, 1965). The dye is absorbed to native DNA as a monomer, but binds to denatured DNA mainly in the form of dimers. Since the absorption and fluorescence characteristics of acridine orange depend upon whether it is bound as a monomer or a dimer, the degree of UV-induced denaturation of the DNA can be measured quantitatively.

The intrinsic viscosity and molecular weight of DNA are lowered (Alexander and Moroson, 1960) by very high doses of irradiation (approx. 1×10^6 ergs·mm^2). These changes are more pronounced when the irradiation is carried out in an oxygen atmosphere, suggesting the possibility of peroxide intermediates in the scission of the sugar–phosphate backbone. The decrease in viscosity reflects not only a reduction in the molecular weight but also an increase in the flexibility of the DNA owing to locally denatured regions.

In the following sections we shall examine some of the ways in which UV-irradiated polynucleotides interact with enzymes and how certain photoproducts can affect the ability of these polynucleotides to serve as templates. There are two different ways of looking at the influence of a photoproduct on the ability of DNA to act as a substrate for an enzymatic reaction. Either the enzyme recognizes a photoproduct as a new chemical entity or else the presence of this photoproduct causes a distortion of the double helix and the enzyme recognizes this distortion as a deviation from the normal double helical structure of the DNA.

B. Biological Properties

1. *Repair Systems*

One of the most important areas of research in photobiology is that of enzymatic repair of damaged DNA. Enzymatic repair systems capable of acting on UV-induced lesions in DNA have been found in numerous biological systems. We shall discuss two of the most widely studied repair systems from the viewpoint of their action on isolated DNA.

a. Excision (Dark Repair). One form of enzymatic repair in cells is known as dark repair; it usually results in excision or removal of intact pyrimidine dimers from the DNA (see review by Strauss, 1969). In some cases the excised dimers are found outside the cell. In Chapters 3 and 4 there is a detailed description of dark repair as observed in bacterial and protozoan systems, respectively. We shall only discuss the first step in this repair process, namely, incision or breaking the chain near the lesion. Various workers (Carrier and Setlow, 1970; J. C. Kaplan *et al.*, 1969; Strauss, 1969; Takagi *et al.*, 1968) have shown that an enzyme purified from *Micrococcus luteus* can make a single-strand break in irradiated DNA at the site of a pyrimidine dimer. The enzyme produces no breaks in unirradiated DNA. The following mechanism has been proposed for the *in vitro* action of this incision enzyme, known as UV-endonuclease (Grossman *et al.*, 1969):

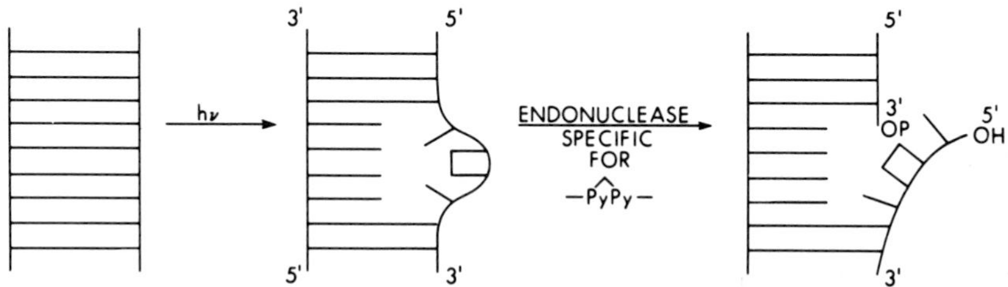

The strand breaks can be observed by use of centrifugation in an alkaline

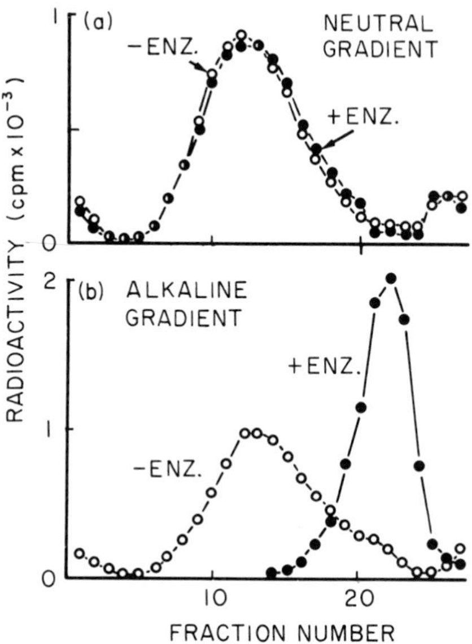

Fig. 16. Sedimentation of UV-irradiated DNA following treatment with an endonuclease from *M. luteus* specific for UV damage. Sedimentation was in an alkaline sucrose density gradient either before or after treatment with the enzyme. Nonirradiated DNA banded at the same position as the irradiated DNA without enzyme. (From Takagi *et al.,* 1968.)

medium. The alkali denatures the DNA, causing strand separation. As illustrated in Fig. 16, there is a marked reduction in the molecular weight of irradiated DNA treated with the endonuclease, indicative of the formation of single-strand breaks. Irradiated DNA not treated with the enzyme does not contain chain breaks. Nor does unirradiated DNA treated with enzyme show any chain breaks. Carrier and Setlow (1970) have shown that these breaks occur in the strand containing the dimer, and that the number of breaks is proportional to the concentration of thymine dimers.

The endonuclease from *M. luteus* is not specific for either the spore photoproduct (R. B. Setlow *et al.,* 1970) or the thymine–cytosine adduct (Patrick and Worthen, 1970) when assayed *in vitro*. According to Carrier and Setlow (1970), the endonuclease has no activity toward irradiated denatured DNA. This result is in conflict with Takagi *et al.* (1968), who found that irradiated denatured DNA was a substrate for the endonuclease. Irradiated poly(A)·poly(U) was able to compete to some degree with irradiated DNA for the enzyme (Carrier and Setlow, 1970), suggesting that the enzyme may bind on ribo- as well as deoxyribopolymers.

Other endonucleases not yet isolated may also be specific for the cytosine–thymine adduct or the spore photoproduct. Experiments *in vivo* with irradiated *Micrococcus radiodurans* (Varghese and Day, 1970) have shown that both the dimer and the cytosine–thymine adduct are excised from irradiated cells. In addition, Stafford and Donnellan (1970) have found that the spore photoproduct formed at low temperatures of irradiation is excised from both *E. coli* and *M. radiodurans* at the same rate. Hence, other enzymes specific for dark repair of photoproducts other than pyrimidine dimers are available *in vivo*.

b. Photoreactivation. J. K. Setlow *et al.* (1965) have examined a yeast extract containing an enzyme that binds to the pyrimidine dimers in irradiated DNA and, in the presence of light in the 300–450 nm range, monomerizes the dimers *in situ*

$$\underline{-\widehat{Y-Y}-} \xrightarrow[\text{light}]{\text{photoreactivating enzyme}} \underline{-YY-}$$

Similar enzymes have been shown to be operative in a wide variety of systems including vertebrate cells (for a review, see Cook, 1970). Cook (1967) actually measured the increase in monomeric thymine following photoreactivation of irradiated DNA and correlated it with the disappearance of thymine dimers.

Strenuous efforts have been made by Muhammed (1966) to purify the photoreactivating enzyme and study the mechanism by which it monomerizes dimers. He encountered several difficulties in trying to prepare pure preparations of the enzyme. First of all, the amount of enzyme present in yeast is small [estimated by Harm and Rupert (1968a) to be 1 part in 1–2 $\times$ 10^5 dry protein] and the overall yield from the purification procedure is low (5–6%). Second, the enzyme in the highly purified state is stable only when complexed with irradiated DNA. An absorbance measurement of the enzyme purified 3000-fold and complexed with DNA failed to show any of the peaks corresponding to those found in the action spectrum for photoreactivation (Fig. 17). At a concentration of photoreactivating enzyme sufficient to bind all the dimers in DNA, a single flash of intense light monomerizes nearly all of the dimers (Harm and Rupert, 1968b). This result indicates that the rate-limiting step in photoreactivation is the time required for the enzyme to move from one lesion to another. A very high degree of photoreactivation occurs even in a glycerol glass at temperatures below 0°C (Harm, 1969) provided the enzyme is allowed to first form a complex with the dimer at higher temperatures. Photoreactivation is considerably reduced at —78°C and is very small at —196°C.

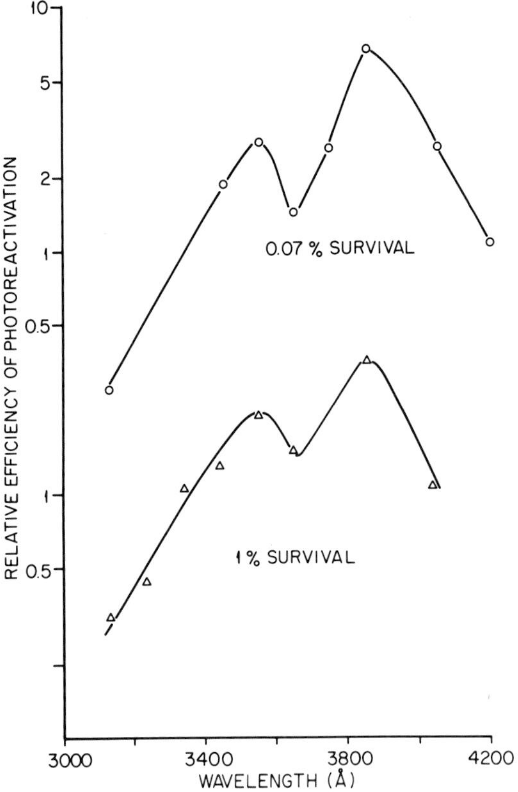

Fig. 17. Action spectrum for photoreactivation of *H. influenzae* transforming DNA with yeast extract. (From J. K. Setlow and Boling, 1963.)

Studies on synthetic polynucleotides have shown that the photoreactivating enzyme is specific for dimers in polydeoxyribonucleotides. Poly(dT), poly(dC), and poly(dU), when irradiated, serve as substrates for the enzyme (J. K. Setlow *et al.*, 1965; J. K. Setlow and Bollum, 1968), but irradiated poly(rC) and poly(rU) (Rupert, 1964), and poly(rT) (J. K. Setlow, 1970) are not substrates. The enzyme is capable of working on single-stranded as well as double-stranded systems. J. K. Setlow and Bollum (1968) measured the ability of the enzyme to monomerize dimers in a series of thymidyl oligonucleotides. They found that the minimum size of the substrate is about nine residues long. According to J. K. Setlow (1966b), thymine–thymine dimers are monomerized ten times faster than cytosine–cytosine dimers and twice as fast as thymine–cytosine dimers. The photoreactivating enzyme appears to monomerize only those pyrimidine dimers commonly found in irradiated native

DNA's. Ben-Hur and Ben-Ishai (1968), for example, found that the photoreactivating enzyme had essentially no effect on the *trans–syn* thymine dimer $\widehat{\mathrm{TT}}_2$. Since the *cis–syn* thymine dimer $\widehat{\mathrm{TT}}_1$, as well as the $\widehat{\mathrm{CC}}$ and $\widehat{\mathrm{CT}}$ dimers formed in DNA, are photoreactivable, the failure to monomerize *trans–syn* dimers supports the view that both the $\widehat{\mathrm{CC}}$ and $\widehat{\mathrm{CT}}$ dimers in DNA are of the *cis–syn* type. Neither the cytosine–thymine adduct, which is a precursor for PO-T (Patrick, 1970), nor the spore photoproduct (Rahn and Hosszu, 1968) is substrate for the photoreactivating enzyme.

Patrick and Worthen (1970) demonstrated that after maximum exposure of irradiated DNA to the endonuclease, the photoreactivating enzyme was no longer able to efficiently monomerize the dimers even though they remained *in situ*. Since the chain break made by the endonuclease is next to a dimer, this result suggests that the photoreactivating enzyme is not able to function on a dimer located at or near the end of a strand.

2. *Transformation*

The loss of DNA-transforming ability after UV-irradiation is an important biological assay for the presence of UV-induced lesions. The transformation assay is based on the observation that cells lacking a genetic marker, such as resistance to streptomycin, can be transformed into resistant cells upon exposure to DNA isolated from a resistant strain. This acquired resistance is due to the incorporation of the DNA from the drug-resistant strain into the genome of the drug-sensitive cells. Irradiation of transforming DNA before it is added to the cells decreases its ability to transform. The transformation assay is unique because the doses of irradiation necessary to achieve an observable decrease in transforming ability are much greater than the doses normally used to inactivate microorganisms. Hence, the photoproduct yields are greater and one can more easily correlate biological inactivation with the formation of photoproducts. R. B. Setlow and Setlow (1962) originally used short-wavelength reversal to demonstrate the biological role played by pyrimidine dimers in the UV inactivation of transforming DNA. They found that a considerable portion of the UV inactivation caused by irradiation at 280 nm could be reversed by further irradiation at 239 nm (see Fig. 2 in Chapter 3). This reactivation was shown to be correlated with the monomerization of pyrimidine dimers. Photoreactivation of irradiated transforming DNA with the photoreactivating enzyme also results in both monomerization of the dimers and restoration of biological activity, with more than 90% of the activity being restored (J. K. Setlow, 1971).

High doses of UV prevent the incorporation of transforming DNA into the host DNA. Muhammed and Setlow (1970) measured the UV-induced decrease in integration of transforming DNA into the DNA of the recipient cells. They found that the $1/e$ dose for inhibition of integration was sufficient to produce dimers that were on the average separated by about 400 nucleotides on a single strand. This length is the same as that shown to be the minimal length for integration of DNA into the genome (Guild *et al.*, 1968).

3. *Enzymatic Hydrolysis*

Exonuclease enzymes that cleave the phosphodiester bonds by sequential action presumably start at one end of the DNA strand and proceed stepwise to the other end, removing one base at a time. R. B. Setlow *et al.* (1964), by comparing the rate of enzymatic hydrolysis of irradiated and nonirradiated DNA, found that the rate of hydrolysis with either venom phosphodiesterase or *E. coli* phosphodiesterase was much less for irradiated DNA than for unirradiated DNA. Using excess enzyme, they showed that the irradiated DNA was enzymatically degraded to monomers plus trinucleotides of the form $N\text{—}\overset{\frown}{Y\text{—}Y}$ where N is any base and $\overset{\frown}{Y\text{—}Y}$ a pyrimidine dimer. Presumably, when the enzyme reaches the dimer it cannot split the phosphodiester bond between the two pyrimidines or between the dimer and the adjacent base. So it works slowly in this region until it begins hydrolysis beyond the dimer. The presence of these "nuclease-resistant sequences" in UV-irradiated DNA greatly impedes the overall rate of enzymatic hydrolysis. The number of nuclease-resistant sequences decreases after treatment of the irradiated DNA with photoreactivating enzyme (R. B. Setlow *et al.*, 1964), a treatment that specifically splits pyrimidine dimers.

4. *Transcription*

DNA can be synthesized *in vitro* in an enzymatic polymerizing system consisting of a polymerase obtained from calf thymus glands and a template consisting of denatured DNA (Bollum, 1959, 1960).

$$\text{Template DNA plus}\ \begin{matrix}\text{dATP}\\ \text{dGTP}\\ \text{dCTP}\\ \text{dTTP}\end{matrix}\ \xrightarrow{\text{DNA polymerase}}\ \text{newly synthesized DNA}$$

Bollum and Setlow (1963) irradiated DNA with UV and studied its ability to serve as a template. They found that the template efficiency decreased with increasing doses of irradiation. They measured as a function of dose the rate of incorporation of nucleotide triphosphates into newly synthesized DNA and found that the rate of incorporation of dATP is more affected by dose than that of dGTP. This is to be expected if the major DNA photoproduct, the thymine dimer, cannot serve as a template for incorporation of ATP. The template activity of the irradiated DNA is partially restored when the DNA is subjected to short-wavelength reversal (i.e., upon splitting pyrimidine dimers). Finally, Bollum and Setlow estimated that 1 thymine dimer per 200 nucleotides decreases the template efficiency by a factor of two. The following model for the effect of a thymine dimer on the rate of DNA synthesis was proposed:

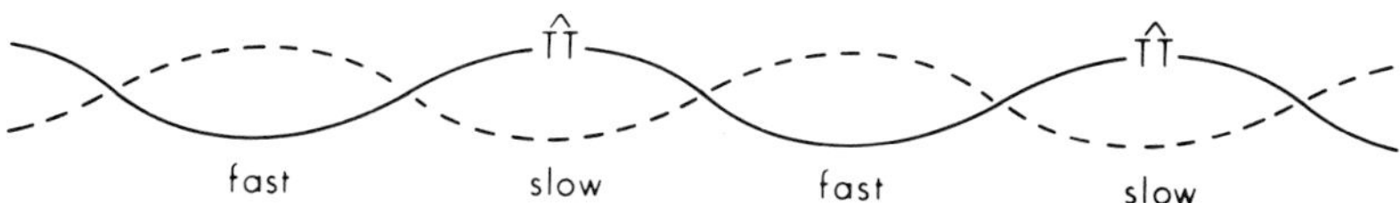

As with the nuclease reaction, the enzyme has difficulty functioning when it reaches a dimer, and the rate of synthesis slows down until some other means is found of getting around the block. The enzyme may add bases in a random manner in the vicinity of the dimer until it returns to the template; or, *in vivo,* it may simply delete some bases and leave a gap in the newly synthesized chain (Rupp and Howard-Flanders, 1968).

VI. Miscellaneous Topics

A. Bromouracil Substitution

When the base analog 5-bromouracil is substituted for thymine in the DNA of various biological systems, the systems become much more sensitive to UV-irradiation. The origin of this increased sensitivity has recently been examined by Lion (1968, 1970) and Hutchinson and Hales (1970). These workers found that UV-irradiation of BrUra-labeled DNA resulted in the random production of single-strand breaks in the DNA. Hutchinson and Hales (1970) detected these breaks by first denaturing the irradiated DNA with either formamide or alkali and then determining, by zone sedimentation, the molecular weight distribution as a function of dose. They rule out the possibility that chain breaks originate only upon alkaline treatment of the irradiated DNA. Irradiation of hybrid DNA containing bromouracil on one strand only indicated that the breaks are made exclusively in the strand containing the bromouracil (Hutchinson and Hales, 1970). Hotz and Walser (1970), however, found both single- and double-strand breaks in BrUra-labeled DNA isolated from irradiated phage. This result suggests that breaks may also occur in the strand opposite the one containing bromouracil. The rate of chain breakage in Hutchinson and Hales' experiment was estimated as 0.007 break per 10^6 daltons$\cdot$ergs$^{-1}\cdot$mm^{-2} for 254 nm irradiation. This rate is two orders of magnitude greater than the rate of breakage in DNA not containing bromouracil.

The enhancement of chain breaks in BrUra-labeled DNA is in part due to the photolability of the bromine atom attached to the ring. Wacker and co-workers (1963, 1964) showed that irradiation of BrUra-labeled DNA led to debromination and formation of uracil

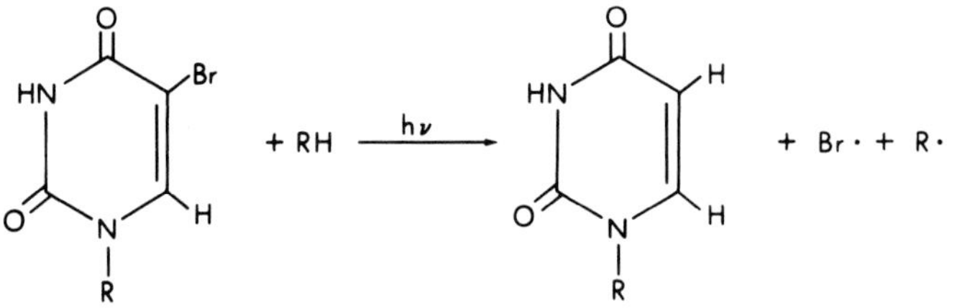

presumably via a free radical intermediate at the C-5 position. Debromination did not occur when BrUra alone was irradiated. Furthermore, the rate of debromination in BrUra-labeled DNA as measured by the appearance of uracil depended upon the state of the DNA. For a given dose of UV, the yield in native DNA was 29% uracil as compared with 13% in denatured DNA and 4.4% in apurinic acid.

Hotz and Reuschl (1967) proposed that the hydrogen atom necessary for the formation of uracil in BrUra-labeled DNA is abstracted from a deoxyribose moiety. This proposal was based on their observations that the deoxyribose of BrUra-labeled DNA was destroyed upon irradiation. Cysteamine, which markedly reduces the UV sensitivity of BrUra-labeled phage, also reduced the rate at which the deoxyribose was destroyed. However, cysteamine has no effect on the rate of destruction of 5-bromouracil in DNA (Lion, 1965).

Support for the mechanisms of Hotz and Reuschl comes from the electron spin resonance studies of Köhnlein and Hutchinson (1969), who showed that irradiation of BrUra-labeled DNA leads to a free radical on the sugar. These workers inspected molecular models of DNA in the

B form and found that the bromine atom in BrUra-labeled DNA is in van der Waals' contact with the sugar attached to the base on the 5′ side of the bromodeoxyuridine nucleotide. They proposed that a hydrogen atom was abstracted from this sugar moiety upon irradiation, and that the abstraction could then lead to a chain break in the phosphodiester backbone. According to this mechanism, the uridine residue left after debromination is on the 3′ side of the chain break.

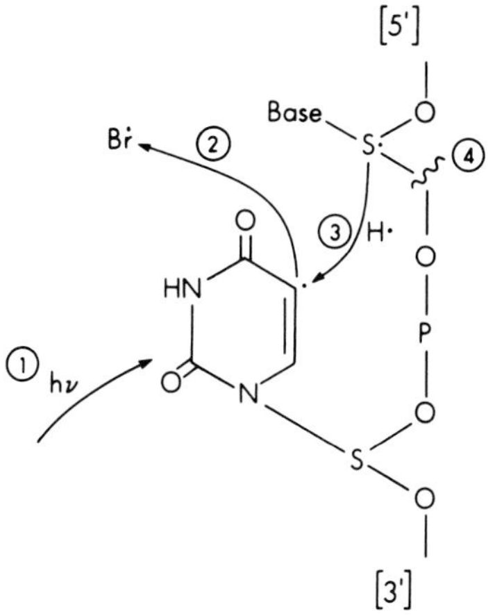

B. Photosensitization

Recently, considerable interest has been generated in the use of triplet sensitizers to form photoproducts in biological systems. This interest was stimulated by the finding of Lamola and Yamane (1967) that irradiation of DNA in the presence of acetophenone with wavelengths greater than 300 nm yields thymine dimers ($\widehat{TT}_1$) as the principal photoproduct. As shown in Table VIII, the rates of formation of the other pyrimidine dimers are much smaller relative to $\widehat{TT}_1$. Photosensitization is thus useful for photobiological studies, since the interpretation of the survival data is simplified by the presence of essentially one photoproduct. A note of caution should be added, however. Under some conditions either addition of the sensitizer to the DNA may take place or the excited sensitizer may induce chain breaks and cross-links in the DNA. These aspects of the problem should always be examined first. Lamola (1969) estimates that not more than one molecule of acetophenone becomes bound to the DNA for every 60 thymine dimers produced.

The mechanism for sensitization involves the absorption of light by a

TABLE VIII

Relative Rates of Photoproduct Formation Obtained upon Acetophenone Sensitization of DNA[a]

Photoproduct	Relative rates of formation
$\widehat{\mathrm{TT}}_1$	1.0
$\widehat{\mathrm{CT}}$	0.03
TH_2	0.02
$\widehat{\mathrm{CC}}$	<0.0025
$\mathrm{C \cdot H_2O}$	<0.003[b]

[a] From Lamola, 1969.
[b] Lamola, 1971.

sensitizer with a very efficient intersystem crossover that converts most of the energy to the triple state. Then, by collision between the sensitizer and the DNA, this triplet energy is transferred to the bases in DNA with lower-lying triplet levels than the organic sensitizer. For acetophenone, only thymine has a lower triplet in DNA (Table IX). With acetone, however, the triplets of all of the other bases can be sensitized because the acetone has a higher-lying triplet state.

Because the radiation wavelengths employed in sensitization ($\lambda > 300$ nm) are not absorbed by the dimer, no photoreversal of the dimers takes place during irradiation. Hence, dimer yields that approach the maximum number of dimers possible are obtained. Sensitization of *E. coli* DNA, for example, results in approximately 40% of the thymine being converted to thymine dimers. This value is close to the percentage of thymine present in DNA as part of a —T—T— sequence (approx. 50%). However, a 50% dimer yield is impossible because of sequences of the form —T—T—T—, in which only two of the thymines can dimerize.

TABLE IX

Extrapolated Values of 0-0 Triplet Energies Determined from the Emission Spectra of the Mononucleotides for A, T, G, and C[a]

Chromophore	$E_{00}(\mathrm{cm}^{-1})$
Acetone	28,200
C	27,900
G	27,250
A	26,700
Acetophenone	26,500
T	26,300

[a] From Lamola *et al.*, 1967.

C. Exogenous Chemicals

The complexing of DNA with certain exogenous chemicals may in some cases affect the photochemistry. The following discussion deals with two different kinds of molecules (metal ions and acridine dyes) that bind to DNA and affect the rate of formation of thymine dimers in DNA.

1. *Metal Ions*

Mercuric ions bind very strongly to the bases in DNA, displacing ring protons and forming cross-links between the strands (Nandi *et al.*, 1965; Yamane and Davidson, 1961). This binding is reversed upon addition of an anion, such as Cl^- or CN^-, that complexes with the mercuric ion. When all the binding sites in DNA have been saturated with mercuric ions, the rate of dimerization is markedly reduced (Fig. 18). Furthermore, when mercuric ions are added to a sample of extensively irradiated DNA, the dimers are reversed upon further irradiation (Fig. 18); i.e., mercuric ions shift the photosteady equilibrium in favor of the monomer. In contrast, mercuric ions bound to poly(dT) have no effect on the rate of photodimerization.

These results for DNA have been interpreted (Rahn *et al.*, 1970) by assuming that mercurated bases other than thymine act as low-lying energy traps. The result is that much of the energy absorbed by thymine is lost via energy transfer to these traps, and less energy is available for dimer formation. Transfer may occur at both the singlet and triplet levels.

B. M. Sutherland and Sutherland (1969b) studied the effects of cupric

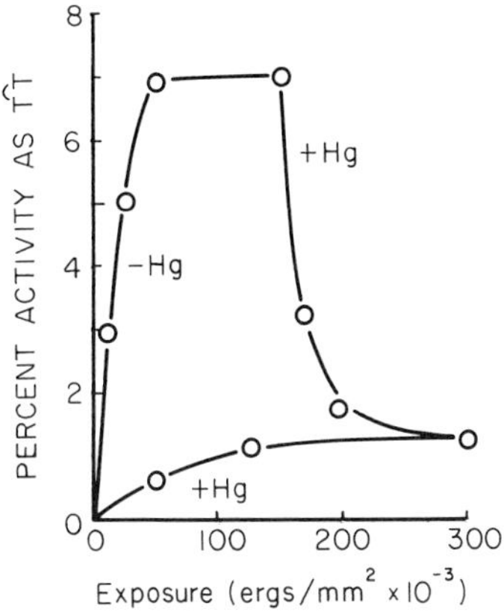

Fig. 18. Influence of mercuration on the formation and reversal of thymine dimers in *E. coli* DNA. Irradiation was at 254 nm. (From Rahn *et al.*, 1970.)

ions on the formation of pyrimidine dimers in DNA. They found that the binding of cupric ions to the bases themselves occurred only in nonhydrogen bonded regions of the DNA and that this binding resulted in a reduction of the yields of thymine dimer by a factor of approximately four. They attributed this quenching to the metal ion either holding the bases apart or else lowering the energy levels of the other bases with respect to thymine.

Recent studies by Rahn and Landry (1971) have shown that the binding of argentous ions to DNA results in up to a 28-fold enhancement in the rate of thymine dimerization. The reason for this enhancement is not yet understood although it may very well reflect a large increase in the triplet state population due to the heavy atom effect. B. M. Sutherland and Sutherland (1969b) also reported that silver ions increased pyrimidine dimer yields in DNA.

2. *Acridine Dyes*

Several workers have shown that the presence of acridine dyes (e.g., proflavine and acridine orange) during UV-irradiation of DNA decreases the thymine dimer yield (Beukers, 1965; R. B. Setlow and Carrier, 1967; B. M. Sutherland and Sutherland, 1969a; J. C. Sutherland and Sutherland, 1970). These dyes bind to DNA by intercalating between adjacent base pairs. Weill and Calvin (1963) showed that acridine dyes, when bound to DNA, have their fluorescence enhanced because of energy transfer at the singlet level from the DNA bases to the dyes. They found that acridine orange, for example, accepts energy from 10 to 20 base pairs. B. M. Sutherland and Sutherland (1969a) similarly proposed that the dimer quenching observed with acridine orange could be explained by singlet transfer from 10 to 14 base pairs to a single dye molecule. Since proflavine only inhibits the rate of formation of dimers and not their reversal (R. B. Setlow and Carrier, 1967), it follows that the steady-state concentration of dimers is altered by its presence. Hence thymine dimers formed in the absence of proflavine are reversed upon irradiation in its presence. It has been shown (J. K. Setlow and Setlow, 1967) that the biological inactivation of UV-irradiated transforming DNA may be similarly reversed by further irradiation in the presence of this dye.

D. Reductive Assay

Miller and Cerutti (1968) have shown that the photohydrate of cytidine is reduced with sodium borohydride as follows

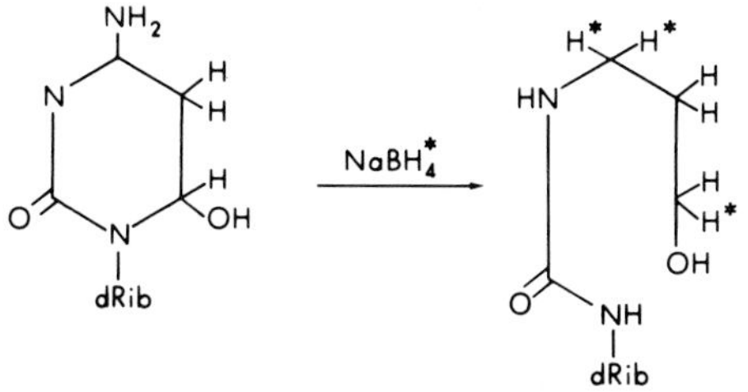

By using sodium borotritiide, one can obtain, after mild acid hydrolysis, the tritiated(*) compound $NH_2C(=O)—NH—CH_2{}^{**}CH_2CH_2{}^{*}OH$, which can then be isolated by ion-exchange and thin-layer chromatography.

Cerutti *et al.* (1969) have used this technique to measure the rate of formation of cytosine photohydrates in irradiated DNA. Under mild acid hydrolysis, the cytosine hydrate is selectively removed from the polymer without any interference from the pyrimidine photodimers. Presumably, under more vigorous hydrolysis conditions degradation products derived from the pyrimidine dimers may also be obtained. Kunieda and Witkop (1967) have demonstrated that the *cis–syn* thymine dimer undergoes the following reaction with sodium borohydride

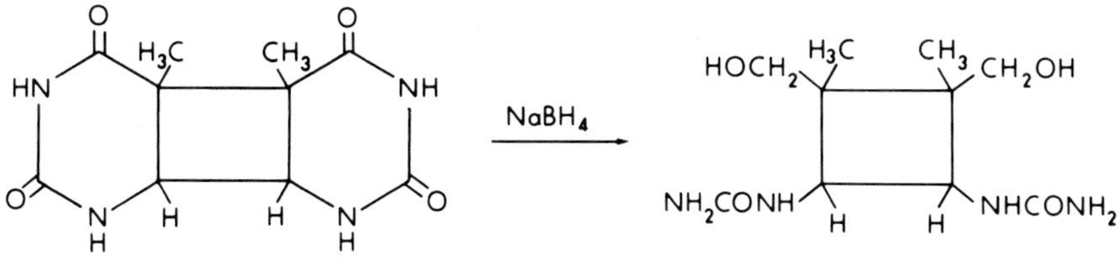

so it may be possible to label pyrimidine dimers in polynucleotides with radioactive material and assay for their presence. This form of assay should prove very useful in situations where one is unable to obtain radioactively labeled DNA. It also provides a method for trapping photoproducts such as hydrates that are short-lived and decay before they can be examined by more conventional procedures.

VII. Summary

In the ten years since thymine dimers were first isolated from irradiated DNA, the field of photobiology has become increasingly more sophisticated. With time, more photoproducts have been discovered and more discriminating methods of measuring these photoproducts have emerged. Most important, the study of repair systems that operate specifically on certain photoproducts has assumed a central role in molecular photobiology. Armed with a knowledge of the relative rates of both photoproduct formation and repair, the invesitgator is in a good position to quantitatively interpret the complex events that follow the irradiation of a cell. It is important to realize that much of the knowledge of photoproduct formation in DNA is based on information obtained from irradiated DNA that has undergone hydrolytic degradation. Hence, acid-labile photoproducts of possible biological significance may be present before hydrolysis but are never isolated because of alterations during the isolation procedure. A fruitful area of future investigation, therefore, would be one connected with the analysis of photoproducts in intact DNA. It is also important to remember that the doses needed to obtain easily measurable quantities of photoproducts are usually much larger than the doses used in the normal biological experiment.

We have also seen that both the nature and the rate of formation of a wide variety of photoproducts in DNA are sensitive to the state of the DNA during irradiation. Because some conditions favor certain photoproducts over others, the photochemical properties of DNA reflect changes in the environment and possibly the secondary structure of the DNA. In this sense, photoproducts can be looked upon as probes of the structure of DNA. Therefore, the study of the photochemistry of DNA is of interest not only to photobiologists but also to those concerned primarily with the structure of DNA.

Finally, the study of the enzymatic repair of photoproducts other than pyrimidine dimers should prove useful in elucidating the mechanism by which repair enzymes recognize specific lesions in the DNA. Undoubtedly, the future will see the isolation of a variety of enzymes intimately involved in some phase of the repair process. The task will be to characterize these enzymes on the basis of their substrate requirements. In particular, it will be desirable to know whether an enzyme such as a UV-specific endonuclease functions on one type of photoproduct but not on another. It is hoped that such studies will help us understand more about recognition phenomena and enzymatic catalysis in biopolymers.

Acknowledgments

I wish to express my appreciation to R. B. Setlow, J. S. Cook, and W. L. Carrier for constructive comments and to Mrs. N. P. Hair, Mr. J. R. Gilbert, Mrs. N. G. Crowe, and Mrs. A. L. Skeel for their help in preparation of the manuscript.

Research was sponsored by the United States Atomic Energy Commission under contract with the Union Carbide Corporation.

References

Adman, E., Jensen, L. H., and Gordon, M. P. (1968). *Abstr., Amer. Crystallogr. Ass. Summer Meet., 1968* p. 103.

Alexander, A., and Moroson, H. (1960). *Nature (London)* **185,** 678.

Baranowska, J., and Shugar, D. (1960). *Acta Biochim. Polon.* **7,** 505.

Becquerel, P. (1910). *C. R. Acad. Sci.* **151,** 86.

Ben-Hur, E., and Ben-Ishai, R. (1968). *Biochim. Biophys. Acta* **166,** 9.

Bersohn, R., and Isenberg, I. (1964). *J. Chem. Phys.* **40,** 3175.

Beukers, R. (1965). *Photochem. Photobiol.* **4,** 935.

Beukers, R., and Berends, W. (1960). *Biochim. Biophys. Acta* **41,** 550.

Blackburn, G. M., and Davies, R. J. H. (1966). *Biochem. Biophys. Res. Commun.* **22,** 709.

Bollum, F. J. (1959). *J. Biol. Chem.* **234,** 2733.

Bollum, F. J. (1960). *J. Biol. Chem.* **235,** 2399.

Bollum, F. J., and Setlow, R. B. (1963). *Biochim. Biophys. Acta* **68,** 599.

Bujard, H. (1970). *J. Mol. Biol.* **49,** 125.

Burr, J. G. (1968). *In* "Advances in Photochemistry" (W. A. Noyes, Jr., G. S. Hammond, and J. N. Pitts, Jr., eds.), Vol. 6, p. 193. Interscience Publishers (Wiley), New York.

Carrier, W. L., and Setlow, R. B. (1970). *J. Bacteriol.* **102,** 178.

Carrier, W. L., and Setlow, R. B. (1971). *Methods Enzymol., Part D* 21, 230.

Cerutti, P. A., Miller, N., and Pleiss, M. (1969). *Abstr. 158th Nat. Meet., Amer. Chem. Soc.* No. 18.

Cook, J. S. (1967). *Photochem. Photobiol.* **6,** 97.

Cook, J. S. (1970). *In* "Photophysiology" (A. C. Giese, ed.), Vol. 5, pp. 191–233. Academic Press, New York.

Davis, S. L., and Tinoco, I. (1966). *Nature (London)* **210,** 1286.

DeBoer, G., and Johns, H. E. (1970). *Biochim. Biophys. Acta* **204,** 18.

Deering, R. A., and Setlow, R. B. (1963). *Biochim. Biophys. Acta* **68,** 526.

Dexter, D. L. (1953). *J. Chem. Phys.* **21,** 836.

Donnellan, J. E. (1971). Private communication.

Donnellan, J. E., Jr., and Setlow, R. B. (1965). *Science* **149,** 308.

Donnellan, J. E., Jr., Hosszu, J. L., Rahn, R. O., and Stafford, R. S. (1968). *Nature (London)* **219,** 964.

Drake, D., and Setlow, R. B. (1962). Unpublished results.

Eisinger, J. (1968). *Photochem. Photobiol.* **7,** 597.

Eisinger, J., and Lamola, A. A. (1967). *Biochem. Biophys. Res. Commun.* **28,** 558.

Eisinger, J., and Shulman, R. G. (1967). *Proc. Nat. Acad. Sci. U. S.* **58,** 895.

Falk, M., Hartman, K. A., Jr., and Lord, R. C. (1963). *J. Amer. Chem. Soc.* **85,** 391.

Fluke, D. J. (1956). *Radiat. Res.* **4,** 193.

Förster, T. (1959). *Discuss. Faraday Soc.* **27,** 7.

Fraenkel, G., and Wulff, D. L. (1961). *Biochim. Biophys. Acta* **51,** 332.

Füchtbauer, W., and Mazur, P. (1966). *Photochem. Photobiol.* **5,** 323.

Gates, F. L. (1930). *J. Gen. Physiol.* **14,** 31.

Gegiou, S., Muszkat, K. A., and Fischer, E. (1968). *J. Amer. Chem. Soc.* **90,** 12.

Giese, A. C. (1962). "Cell Physiology," p. 200. Saunders, Philadelphia, Pennsylvania.

Giese, A. C. (1968). *Photochem. Photobiol.* **8,** 527.

Glišin, V. R., and Doty, P. (1967). *Biochim. Biophys. Acta* **142,** 314.

Greenstock, C. L., and Johns, H. E. (1968). *Biochem. Biophys. Res. Commun.* **30,** 21.

Grossman, L., Levine, S. S., and Allison, W. S. (1961). *J. Mol. Biol.* **3,** 47.
Grossman, L., and Rogers, E. (1968). *Biochem. Biophys. Res. Commun.* **33,** 975.
Grossman, L., Kaplan, J., Kushner, S., and Mahler, I. (1969). *Ann. Ist. Super. Sanita* **5,** Spec. Suppl. 1, 318–333.
Gueron, M., and Shulman, R. G. (1968). *Annu. Rev. Biochem.* **37,** 571.
Guild, W. R., Cato, A., Jr., and Lacks, S. (1968). *Cold Spring Harbor Symp. Quant. Biol.* **33,** 643.
Harm, H. (1969). *Biophys. J.* **9,** A-203.
Harm, H., and Rupert, C. S. (1968a). *Abstr. Int. Congr. Photobiol., 5th, 1968* p. 47.
Harm, H., and Rupert, C. S. (1968b). *Mutat. Res.* **6,** 355.
Haug, A., and Sauerbier, W. (1965). *Photochem. Photobiol.* **4,** 555.
Herbert, M. A., LeBlanc, J. C., Weinblum, D., and Johns, H. E. (1969). *Photochem. Photobiol.* **9,** 33.
Hill, R. F., and Rossi, H. H. (1954). *Radiat. Res.* **1,** 282.
Hosszu, J. L., and Rahn, R. O. (1967). *Biochem. Biophys. Res. Commun.* **29,** 327.
Hotz, G., and Reuschl, H. (1967). *Mol. Gen. Genet.* **99,** 5.
Hotz, G., and Walser, R. (1970). *Photochem. Photobiol.* **12,** 207.
Hutchinson, F., and Hales, H. (1970). *J. Mol. Biol.* **50,** 59.
Jagger, J. (1967). "Introduction to Radiation in UV Photobiology." Prentice-Hall, Englewood Cliffs, New Jersey.
Janion, C., and Shugar, D. (1967). *Acta Biochim. Pol.* **14,** 293.
Jellinek, T., and Johns, R. B. (1970). *Photochem. Photobiol.* **11,** 349.
Johns, H. E. (1968). *Photochem. Photobiol.* **8,** 547.
Johns, H. E., Pearson, M. L., LeBlanc, J. C., and Helleiner, C. W. (1964). *J. Mol. Biol.* **9,** 503.
Johns, H. E., LeBlanc, J. C., and Freeman, K. B. (1965). *J. Mol. Biol.* **13,** 849.
Kaplan, J. C., Jushner, S. R., and Grossman, L. (1969). *Proc. Nat. Acad. Sci. U. S.* **63,** 144.
Kaplan, K. (1955). *Naturwissenschaften* **42,** 466.
Karle, I. L., Wang, S. Y., and Varghese, A. J. (1969). *Science* **164,** 183.
Köhnlein, W., and Hutchinson, F. (1969). *Radiat. Res.* **39,** 745.
Kunieda, T., and Witkop, B. (1967). *J. Amer. Chem. Soc.* **89,** 4232.
Lamola, A. A. (1969). *Photochem. Photobiol.* **9,** 291.
Lamola, A. A. (1971). Personnel communication.
Lamola, A. A., and Mittal, J. P. (1966). *Science* **154,** 1560.
Lamola, A. A., and Yamane, T. (1967). *Proc. Nat. Acad. Sci. U. S.* **58,** 443.
Lamola, A. A., Gueron, M., Yamane, T., Eisinger, J., and Shulman, R. G. (1967). *J. Chem. Phys.* **47,** 2210.
Levine, M., and Cox, E. (1963). *Radiat. Res.* **18,** 213.
Lion, M. B. (1965). *Radiat. Res.* **25,** 211.
Lion, M. B. (1968). *Biochim. Biophys. Acta* **155,** 505.
Lion, M. B. (1970). *Biochim. Biophys. Acta* **209,** 24.
Luyet, B., and Rasmussen, D. (1968). *Biodynamica* **10,** 167.
McLaren, A. D., and Shugar, D. (1964). "Photochemistry of Proteins and Nucleic Acids." Macmillan, New York.
Marmur, J., and Grossman, L. (1961). *Proc. Nat. Aacd. Sci. U. S.* **47,** 778.
Marmur, J., Anderson, W. F., Matthews, L., Berns, K., Gajewska, E., Lane, D., and Doty, P. (1961). *J. Cell. Comp. Physiol.* **58,** Suppl. 1, 33.
Miller, N., and Cerutti, P. (1968). *Proc. Nat. Acad. Sci. U. S.* **59,** 34.
Muel, B., and Malpiece, C. (1969). *Photochem. Photobiol.* **10,** 283.

Muhammed, A. (1966). *J. Biol. Chem.* **241,** 516.
Muhammed, A., and Setlow, J. K. (1970). *J. Bacteriol.* **101,** 444.
Nagata, C., Imamura, A., Tagashira, Y., and Kodama, M. (1965). *J. Theoret. Biol.* **9,** 357.
Nandi, U. S., Wang, J. C., and Davidson, N. (1965). *Biochemistry* **4,** 1687.
Patrick, M. H. (1970). Private communication.
Patrick, M. H. (1970). *Photochem. Photobiol.* **11,** 477.
Patrick, M. H., and Worthen, G. S. (1970). *Biophys. Soc. Annu. Meet. Abstr.* **10,** 190.
Pearson, M. L., and Johns, H. E. (1966). *J. Mol. Biol.* **20,** 215.
Pullman, B. (1965). *In* "Molecular Biophysics" (B. Pullman and M. Weissbluth, eds.), p. 173. Academic Press, New York.
Rahn, R. O. (1970). *In* "Photochemistry of Macromolecules" (R. F. Reinisch, ed.), pp. 15–29. Plenum Press, New York.
Rahn, R. O., and Hosszu, J. L. (1968). *Photochem. Photobiol.* **8,** 53.
Rahn, R. O., and Hosszu, J. L. (1969). *Biochim. Biophys. Acta* **190,** 126.
Rahn, R. O., and Landry, L. C. (1969). *Biophys. J.* **9,** A-269.
Rahn, R. O., and Landry, L. C. (1971). *Biochim. Biophys. Acta* **247,** 197.
Rahn, R. O., Setlow, J. K., and Hosszu, J. L. (1969). *Biophys. J.* **9,** 510.
Rahn, R. O., Battista, M., and Landry, L. C. (1970). *Proc. Nat. Acad. Sci. U. S.* (in press).
Rupert, C. S. (1964). *In* "Photophysiology" (A. C. Giese, ed.), Vol. 2, p. 311. Academic Press, New York.
Rupp, W. D., and Howard-Flanders, P. (1968). *J. Mol. Biol.* **31,** 291.
Setlow, J. K. (1970). Private communication.
Setlow, J. K. (1966a). *Curr. Top. in Radiat. Res.* **2,** 222–248.
Setlow, J. K. (1966b). *Radiat. Res.* **6,** Suppl., 141.
Setlow, J. K. (1971). *Res. Progr. in Org.-Biol. Med. Chem.* **3** (in press).
Setlow, J. K., and Boling, M. E. (1963). *Photochem. Photobiol.* **2,** 471.
Setlow, J. K., and Bollum, F. J. (1968). *Biochim. Biophys. Acta* **157,** 233.
Setlow, J. K., and Setlow, R. B. (1967). *Nature (London)* **213,** 907.
Setlow, J. K., Boling, M. E., and Bollum, F. J. (1965). *Proc. Nat. Acad. Sci. U. S.* **53,** 1430.
Setlow, R. B. (1968a). *Progr. Nucl. Acid Res. Mol. Biol.* **8,** 257.
Setlow, R. B. (1968b). *Photochem. Photobiol.* **7,** 643.
Setlow, R. B., and Carrier, W. L. (1963). *Photochem. Photobiol.* **2,** 49.
Setlow, R. B., and Carrier, W. L. (1966). *J. Mol. Biol.* **17,** 237.
Setlow, R. B., and Carrier, W. L. (1967). *Nature (London)* **213,** 906.
Setlow, R. B., and Carrier, W. L. (1970). *Biophys. Soc. Annu. Meet. Abstr.* **10,** 255a.
Setlow, R. B., Carrier, W. L., and Rahn, R. O. (1970). Unpublished results.
Setlow, R. B., and Doyle, B. (1953). *Biochim. Biophys. Acta* **12,** 508.
Setlow, R. B., and Setlow, J. K. (1962). *Proc. Nat. Acad. Sci. U. S.* **48,** 1250.
Setlow, R. B., Carrier, W. L., and Bollum, F. J. (1964). *Biochim. Biophys. Acta* **91,** 446.
Shapiro, R., and Klein, R. S. (1967). *Biochemistry* **6,** 3576.
Shulman, R. G., and Rahn, R. O. (1966). *J. Chem. Phys.* **45,** 2940.
Sinsheimer, R. L., and Hastings, R. (1949). *Science* **110,** 525.
Smith, K. C. (1963). *Photochem. Photobiol.* **2,** 503.
Smith, K. C. (1964). *In* "Photophysiology" (A. C. Giese, ed.), Vol. 2, pp. 329–388. Academic Press, New York.

Smith, K. C. (1967). *In* "Radiation Research" (G. Silini, ed.), p. 756. Wiley, New York.
Smith, K. C. (1968). *Photochem. Photobiol.* **7,** 651.
Smith, K. C., and Hanawalt, P. C. (1969). "Molecular Photobiology: Inactivation and Recovery" Academic Press, New York.
Smith, K. C., and Meun, D. H. C. (1968). *Biochemistry* **7,** 1033.
Smith, K. C., and O'Leary, M. E. (1967). *Science* **155,** 1024.
Smith, K. C., and Yoshikawa, H. (1966). *Photochem. Photobiol.* **5,** 777.
Snyder, L. C., Shulman, R. G., and Neumann, D. B. (1970). *J. Chem. Phys.* **53,** 256.
Stafford, R. S., and Donnellan, J. E., Jr. (1970). Private communication.
Stafford, R. S., and Donnellan, J. E., Jr. (1968). *Biophys. J.* **9,** 510.
Strauss, B. S. (1969). *Curr. Top. Microbiol. Immunol.* **44,** 1.
Sutherland, B. M., and Sutherland, J. C. (1969a). *Biophys. J.* **9,** 292.
Sutherland, J. C., and Sutherland, B. M. (1970). *Biopolymers* **9,** 639.
Sutherland, B. M., and Sutherland, J. C. (1969b). *Biophys. J.* **9,** 1329.
Sztumpf, E., and Shugar, D. (1962). *Biochim. Biophys. Acta* **61,** 555.
Takagi, Y., Sekiguchi, M., Okubo, S., Nakayama, H., Shimada, K., Yasuda, S., Nishimoto, T., and Yoshihara, H. (1968). *Cold Spring Harbor Symp. Quant. Biol.* **33,** 219.
Trifonov, E. N., Shafranovskaya, N. W., Frank-Kamenetskii, M. D., and Lazurkin, Yu. S. (1968). *Mol. Biol.* **2,** 887.
Varghese, A. J. (1970). *Biochem. Biophys. Res. Commun.* **38,** 484.
Varghese, A. J., and Day, R. S., III. (1970). *Photochem Photobiol.* **11,** 511.
Varghese, A. J.. and Wang, S. Y. (1967). *Science* **156,** 955.
Varghese, A. J., and Wang, S. Y. (1968). *Science* **160,** 186.
Wacker, A. (1963). *Progr. Nucl. Acid Res.* **1,** 369.
Wacker, A., Dellweg, H., Träger, L., Kornhauser, A., Lodemann, E., Türck, G., Selzer, R., Chandra, P., and Ishimoto, M. (1964). *Photochem. Photobiol.* **3,** 369.
Wang, S. Y., and Varghese, A. J. (1967). *Biochem. Biophys. Res. Commun.* **29,** 543.
Webb, S. J. (1965). "Bound Water in Biological Integrity." Thomas, Springfield, Illinois.
Wei, C. H., and Einstein, J. R. (1968). *Abstr. Amer. Crystallogr. Ass. Summer Meet., 1968* p. 102, L-9.
Weill, G., and Calvin, M. (1963). *Biopolymers* **1,** 401.
Weinblum, D. (1967). *Biochem. Biophys. Res. Commun.* **27,** 384.
Weinblum, D., and Johns, H. E. (1966). *Biochim. Biophys. Acta* **114,** 450.
Wierzchowski, K. L., and Shugar, D. (1961). *Acta Biochim. Pol.* **2,** 219.
Wulff, D. L. (1963). *Biophys. J.* **3,** 355.
Yamane, T., and Davidson, N. (1961). *J. Amer. Chem. Soc.* **83,** 2599.
Yamane, T., Wyluda, B. J., and Shulman, R. G. (1967). *Proc. Nat. Acad. Sci. U. S.* **58,** 439.
Zavil'gel'skii, G. B., Borisova, O. F., Minchenkova, L. E., and Minyat, E. E. (1964). *Biokhimiya* **29,** 508.
Zavil'gel'skii, G. B., Minchenkova, L. E., Minyat, E. E., and Savich, A. P. (1965). *Biokhimiya* **30,** 652.

Chapter 2

Methods for Photoinactivation of Viruses

C. W. HIATT

I. Introduction

Studies of virus inactivation with ultraviolet light or photodynamic action have been variously motivated. In some instances, the basic photochemical mechanism has been the subject of interest (Uber, 1941; Fluke and Pollard, 1949; Zelle and Hollaender, 1954; McLaren and Takahashi, 1957; McLaren and Moring-Claesson, 1961). Other investigators have employed photoinactivation as a means to gain insight into the molecular biology of viruses (Bawden and Kleczkowski, 1953; Chessin, 1960). The practical applications in classifying viruses (Latarjet and Wahl, 1945; Baron *et al.*, 1959; Welsh and Adams, 1954; Moore *et al.*, 1962) or in preparing vaccines and diagnostic antigens (Hodes *et al.*, 1940; Habel and Sockrider, 1947; Shaughnessy *et al.*, 1957; Perdrau and Todd, 1933b; Wallis *et al.*, 1963) have also received their share of attention.

All of these courses of investigation draw upon a common set of principles for the design of apparatus, measurement of dose, and interpretation of kinetic data, and it is convenient to consider them together. As a group, ultraviolet (UV) inactivation reactions and the various photooxidative processes known as photodynamic action (PDA) differ from the effects induced by ionizing radiation in two important respects: (a) Their quantum efficiency is low (10^{-4} to 10^{-2}), and (b) they are capable of selective action upon molecules of biological interest. Both of these characteristics result from the fact that the quantum energy of the radiation is too low to cause ionization (i.e., to dislodge an electron from an atom) but has its effect through excitation of certain atoms, or molecules, to energy levels which permits them to engage in chemical reactions that are exceedingly improbable in the ground state, or normal energy level.

The quantum energies associated with various wavelengths of light are

TABLE I
Quantum Energy of the Photon for Various Wavelengths of Light[a]

Wavelength (λ) (nm)	Energy (*E*)	
	Ergs × 10^{-12}	Kcal/mole
253.7	7.82	112.5
260	7.62	109.6
280	7.09	102.0
320	6.23	89.6
440	4.51	64.9
540	3.71	53.4
630	3.18	45.8
700	2.85	41.0

listed in Table I. It may be seen that, even at 253.7 nm, the principal emission wavelength of low-pressure mercury lamps, the quantum energy is only 112.5 kcal/mole, barely intruding upon the range (90–500 kcal/mole) of energies required to ionize neutral atoms and is too low for direct ionization to assume any degree of importance in reactions induced by light of this wavelength.

II. Ultraviolet Light

A. Light Sources

A large proportion of the experiments undertaken in the photobiology of viruses have utilized light from low-pressure mercury vapor lamps which emit most of their light energy at a wavelength of 253.7 nm. As indicated in Table II, more than 95% of the spectral intensity of a 15-W "germicidal" lamp is between 252 and 255 nm. Lamps of this type are virtually monochromatic sources and especially useful for virus inactivation experiments, since their principal emission band is quite close to the maximum (ca. 260 nm) in the absorption spectrum of nucleic acid. As pointed out by Jagger (1961), however, the output of this type of lamp at

TABLE II

Spectral Intensity Perpendicular to Axis of 15-W Low-Pressure Mercury Vapor Lamp[a]

Wavelength band (nm)	Intensity ($\mu W/cm^2$ at 1 m from lamp)
247.0–249.7	0.01
249.7–252.4	0.06
252.4–255.0	16.0
255.0–260.7	0.03
260.7–272.4	0.02
272.4–282.0	0.01
282.0–286.1	0.01
286.1–292.6	0.01
292.6–299.8	0.06
299.8–305.0	0.03
305.0–322.6	0.29
322.6–343.4	0.02
343.4–360.1	0.005
360.1–369.6	0.31
369.6–379.8	0.03
379.8–390.9	0.05

[a] Data from Koller (1939).

253.7 nm becomes a lesser fraction of the total emission energy as the lamp ages in use and may drop to 73% or less after 500 hr.

The General Electric series of low-pressure mercury vapor lamps includes cylindrical tubes in Vycor envelopes ranging in length from 12 to 36 inches and in rated wattage from 8 to 36. These lamps are electrically identical to fluorescent tubes of the same dimensions. The 15-W G15T8 lamp (18 inches long) is an especially convenient one for the experimenter, since one or two such lamps may be mounted in a common type of desk lamp fixture.

A smaller lamp (G4T4), rated at 4 W, is U-shaped, about 5 inches in length, and is mounted on a 4-prong radio type base plug. Figure 1 shows the design of a compact and efficient housing for this lamp.

For determining action spectra or for other experiments requiring the full range of ultraviolet wavelengths, other types of mercury vapor or xenon arc lamps are required. The properties of these sources are discussed by Jagger (1967) in a monograph which also includes some valuable information about other aspects of ultraviolet photobiology.

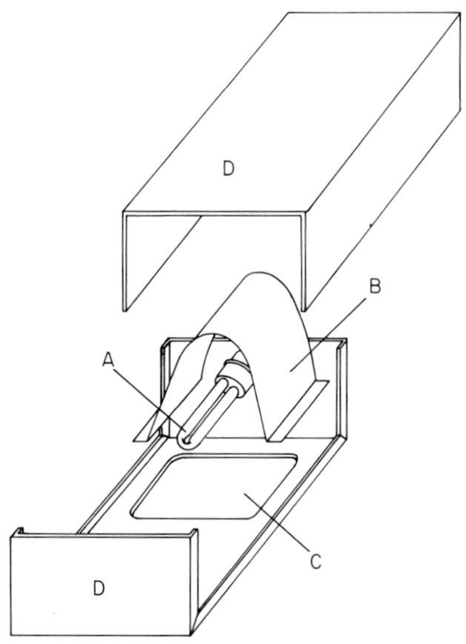

Fig. 1. Exploded view of a parabolic reflector mounting for the General Electric G4T4-1 germicidal lamp. The lamp (A) is mounted in a 4-pronged radio tube socket with the center lines of the two limbs of the envelope at the principal foci of the double paraboloid surface of the polished aluminum reflector (B). The aperture (C) is cut out of the bottom of the aluminum box (D) (Bud CU-3010-A). Not shown in this diagram are the ballast (GE 6G1041) and fluorescent lamp starter (FS-5) required to supply the operating current for the lamp.

B. Light Measurement Techniques

There are three levels of achievement in measuring light intensity in photoinactivation experiments. The first of these is to determine, on a relative basis, the total irradiance at all wavelengths from a given source. One can then expect to reproduce experimental conditions and administer precise graded doses of light. The second level is to be able to describe relative irradiance as a function of wavelength, which then permits the investigator to obtain action spectra and take the initial step in identifying the target molecule or site of inactivation. The ultimate level of attainment is to measure the absolute irradiance at all wavelengths and to record this in the accepted units: microwatts/cm^2 or ergs/mm^2/sec (1 erg mm^{-2} sec^{-1} is equivalent to 10μW cm^{-2}). It is then possible to calculate the number of quanta of light required to achieve a given biological effect and to determine quantum efficiency (sometimes called "quantum yield") in absolute terms. Quantum efficiency is the number of biological units (molecules or particles or infectious units) reacting divided by the number of quanta absorbed. To be able to describe a photobiological process in terms of quantum efficiency is an achievement to be proud of, since it puts the information into proper language for assimilation into the neighboring field of photochemistry.

There is a point of caution to be observed, however, in attempting to express irradiance measurements in absolute terms without laboratory standardization of the measuring instruments. Errors in calibration or notation are easily perpetuated by the busy experimenter who may be most interested in relative effects, after all, and chooses absolute notation only to fulfill a presumed requirement for respectability.

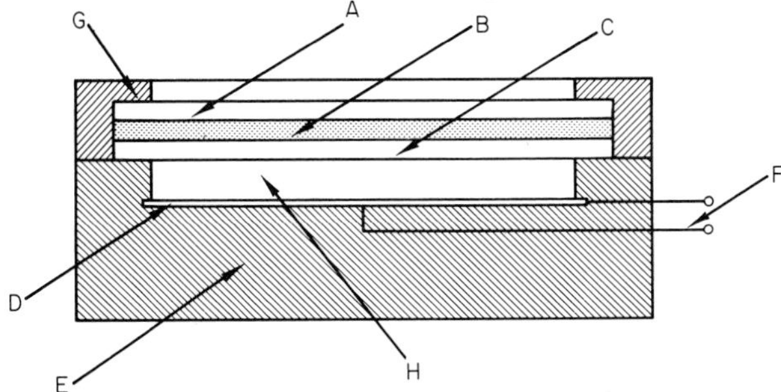

Fig. 2. Schematic cross-section of the Latarjet ultraviolet meter. (A) Corning filter No. 9863, (B) layer of zinc orthosilicate, (C) "Katathermic" filter, (D) selenium layer, (E) housing, (F) leads to microammeter, (G) retaining cover, (H) air space. The current produced by the selenium cell is proportional to the incident intensity of light at a wavelength of 254 nm. (Redrawn from Latarjet *et al.*, 1953.)

For experiments employing a low-pressure mercury arc lamp, such as those previously described, probably the most convenient device for measuring light intensity is the photoelectric meter devised by Latarjet *et al.* (1953) and manufactured for sale on a nonprofit basis by Dr. R. Latarjet, Institut du Radium, 26 rue d'Ulm, Paris V[e], France. The sensing unit, shown in schematic form in Fig. 2, is a selenium photovoltaic cell covered by two filters, between which is a phosphor layer (zinc orthosilicate). The top filter is relatively transparent at 254 nm but cuts off light below 200 nm and above 400 nm in the visible range. The phosphor, excited by the light that passes the top filter, emits visible fluorescence that passes through the bottom filter and energizes the selenium layer below it. The bottom filter shields the selenium cell from infrared light, which may penetrate the top filter. The characteristic spectral responses of the four components of this device operate together to achieve the highly specific sensitivity curve shown in Fig. 3. Output of the selenium cell, measured by a microammeter with an internal resistance of 3000 ohms or less, is directly proportional to the incident intensity at 254 nm, and the cell will maintain calibration for several years if handled with reasonable care.

A similar sensing device described by Jagger (1961) has greater sensitivity but is less specific in its response to 254 nm light. Either device may be used without interference from ordinary room illumination.

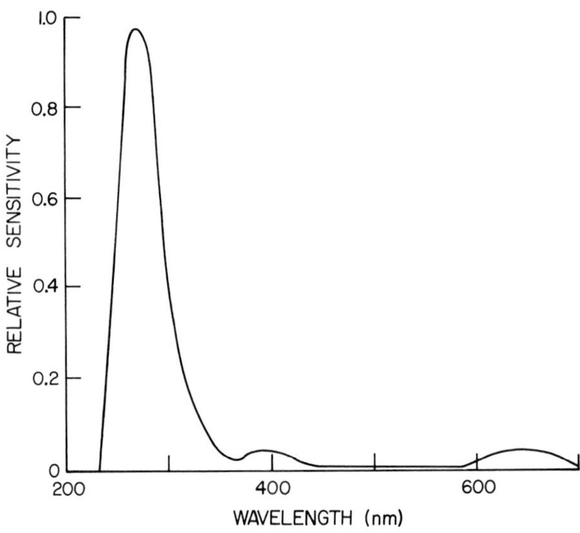

Fig. 3. Spectral sensitivity of the Latarjet ultraviolet meter. (Redrawn from Latarjet *et al.,* 1953.)

C. Calibration

To calibrate meters of the type described, the traditional approach in the United States is to compare the meter response to that of a sensitive thermopile which in turn is calibrated against a secondary radiation standard (a 50-W carbon filament lamp) supplied by the National Bureau of Standards. The detailed procedure is given by Stair and Johnston (1954). Reliably calibrated thermopiles are available commercially (e.g., The Eppley Laboratory, Inc., Newport, Rhode Island) and highly sensitive lock-in voltmeters (such as the Model 131 manufactured by Brower Laboratories, Inc., Waltham, Massachusetts) have done much to simplify the onerous task of calibration.

The experimenter who wishes to avoid the expense and effort required to calibrate his own equipment may purchase the Latarjet meter complete with microammeter and calibration curves or may make suitable arrangements with the Radiometry Section, Atomic and Radiation Physics Division, National Bureau of Standards, to have the calibration done.

D. Determination of Incident Intensity

The measurement of incident intensity is simplified if the surface of the liquid to be irradiated is at right angles to the rays of light and if there is no intervening vessel wall. Then the reference plane of the sensing element is put in the same plane as the surface of the liquid and no correction factor is required. (The slight difference in reflectance between the surface of the liquid and the window of the sensing element is usually negligible.)

An example of an irradiation device of this type is shown in Fig. 4. A measured volume of the liquid to be irradiated is placed in the flat-bottomed cavity in the glass slide. By rotating the crossarm, the sensing element (in this case a thermopile) may be exactly positioned under the light source, permitting the measurement of intensity between samples if desired.

For short time intervals (5 min or less) evaporation of water from the exposed sample is minimal, and microbiological contamination is rarely a problem because of the sterilizing effect of the UV light upon the air above the sample. For some experiments, however, it may be desirable to place quartz cover slips over both the sample chamber and the sensing element.

Another device which has been used with success for irradiating small volumes of liquid is depicted in Fig. 5. Measured volumes are placed in

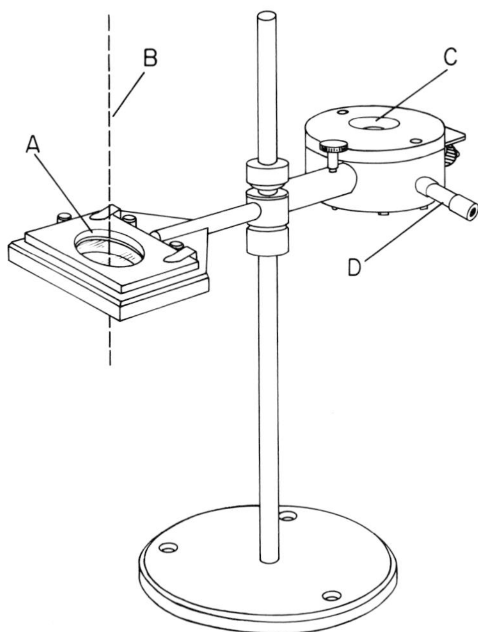

Fig. 4. Irradiation vessel with coplanar sensing element. (A) glass or quartz cavity slide in holder, (B) optical axis, (C) aperture of calibrated thermopile, (D) tube for attaching vacuum pump. (Thermopiles may be used satisfactorily at ambient pressures, without employing the vacuum system.)

each of the four cavities of a quadrant petri dish. The dish on its carriage is rotated slowly under the light source to provide equal dose rates for each of the quadrants. The light source may be quite close above the petri dish, since the rotation of the samples eliminates the effect of nonuniformities in illumination. At timed intervals the irradiation is interrupted by interposing a sheet of opaque material between the light source and the sample. Rotation of the dish is stopped, an aliquot withdrawn, and the irradiation is then resumed. Thus, four samples may be irradiated concurrently, and the withdrawal of aliquots does not affect the layer thickness of the other samples. (This procedure is valid only when the reaction under investigation obeys the reciprocity law, i.e., the progress of the reaction depends upon the total number of light quanta absorbed, without regard to the dose rate or discontinuity of the illumination.)

In neither of the two devices shown has there been any provision for stirring the liquid during irradiation. As our subsequent discussion will indicate, attempts to stir small volumes of liquid to improve the uni-

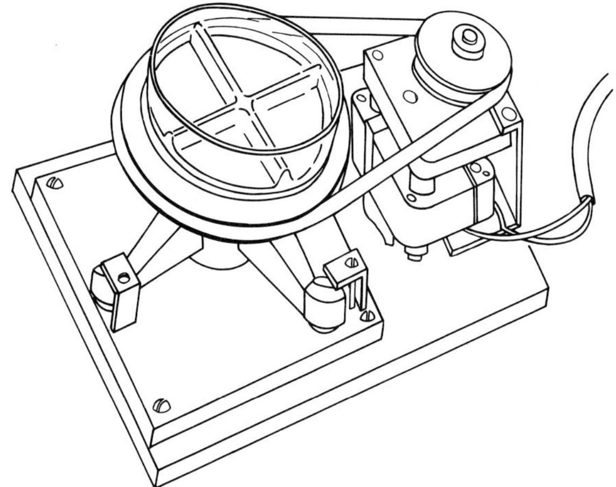

Fig. 5. Device for rotating a quadrant petri dish under a light source to eliminate nonuniformities in illumination.

formity of illumination represent an approach which is less opportune than simply using thin enough layers of liquid so that stirring is unnecessary. Nevertheless, there are instances in which stirring is required to maintain a uniform suspension of cellular debris in unclarified virus preparations. To some extent this may be accomplished by putting a thin layer of the sample in a petri dish and swirling it slowly on an orbital shaking platform under the light source.

Oster and McLaren (1950) used a cuvette with a quartz stirrer to mix suspensions of tobacco mosaic virus during irradiation (see Fig. 6). The necessity of using transparent material for the stirrer points out one of the special requirements of inactivation experiments with microbial systems. Since it is often desirable to follow the course of such reactions to very low survival ratios (10^{-6} or less) it is important to avoid using irradiation vessels in which even minute quantities of the sample are shielded from light.

E. Actinometers

Another approach to the determination of incident intensity is to use a chemical or biological actinometer. An actinometer is a solution which is placed in the sample cell and which, upon irradiation, changes in some

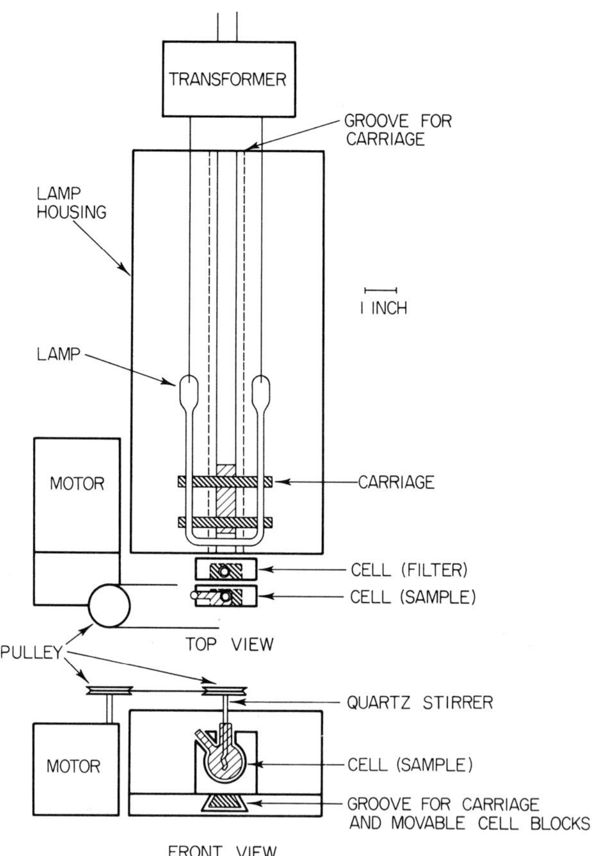

Fig. 6. Irradiation apparatus using quartz stirrer. (From Oster and McLaren, 1950.)

measurable property as a function of the number of light quanta absorbed. The traditional system for this purpose (see Bowen, 1946) is a dilute solution of uranyl oxalate (0.05–0.5%) in an excess of oxalic acid. Upon irradiation the uranyl oxalate decomposes and the remainder is titrated against an acidified standard solution of potassium permanganate. Now in more common use is a solution of potassium ferrioxalate, analyzed by a colorimetric reaction with phenanthroline (Hatchard and Parker, 1956; see also Jagger, 1967).

While in concept these solutions may be used to simulate virus suspensions *in situ* in the sample cell, in practice it has been found preferable to use them at much higher optical densities than would be appropriate for the virus suspension itself. In fact it is desirable to have an optical density high enough to absorb virtually all of the incident light. By finding the quantity of photosensitive substance degraded per unit time (at constant intensity) and with reference to tabulated values for the quantum efficiency of the actinometer at the wavelength concerned, one may estimate the incident intensity in absolute terms without any optical calibration.

Biological actinometers have sometimes been employed. Latarjet *et al.* (1953) proposed using suspensions of T2 coliphage or C16 *Bacillus dysenteriae* phage (strain Y6R) as standards of reference. Either of these viruses is reduced in survival ratio to 10^{-2} by 220 to 230 ergs mm^{-2} of light at 254 nm. The relatively poor precision of phage assays and potential variation in the UV resistance of the standard phage preparation have limited the usefulness of this technique.

F. Estimation of Quantum Dose

The determination of incident intensity is only the first step in ultraviolet dosimetry. The amount of light reaching each of the virus particles is what determines its fate, and the degree of penetration of light into a volume of liquid is therefore of great importance.

The average intensity of light in a layer of liquid L cm thick, when illuminated on one surface at an incident intensity I_0 is

$$\bar{I} = \frac{I_0(1 - e^{-\alpha L})}{\alpha L} \quad (1)$$

in which α is the absorption coefficient (2.203 times the optical density of a 1-cm layer).

If the layer of liquid is supported upon a reflective surface, Eq. (1) becomes

$$\bar{I} = \frac{I_0}{\alpha L}\,(1 - e^{-\alpha L} + Re^{-2\alpha L}(\alpha L - 1)) \quad (2)$$

in which R is the reflection coefficient (cf. Hiatt, 1961). In Table III, which gives $\bar{I}/I_0$ for graded values of reflection coefficient and transmittance ($e^{-\alpha L}$), it may be seen that the relative average light intensity is markedly increased by the use of reflective supporting surfaces, especially for liquid layers of high transmittance.

The most effective reflecting surface for ultraviolet light is a freshly deposited film of evaporated aluminum. Hass (1955) lists reflection coefficients approximating 0.92 for this material over the wavelength range 220 to 400 nm. Silver films deposited in the same manner are relatively

TABLE III

Average Relative Light Intensity ($\bar{I}/I_0$) in a Layer of Liquid Illuminated from One Side Only and Supported by Surfaces of Various Reflection Coefficients

Transmittance ($e^{-\alpha L}$)	Reflection coefficient 0	0.1	0.2	0.3	0.4	0.5	0.6	0.7	0.8	0.9
0.1	0.391	0.395	0.399	0.403	0.406	0.410	0.414	0.418	0.422	0.426
0.2	0.497	0.507	0.517	0.527	0.537	0.547	0.557	0.567	0.577	0.587
0.3	0.581	0.599	0.616	0.634	0.651	0.669	0.686	0.704	0.721	0.738
0.4	0.655	0.681	0.707	0.733	0.760	0.786	0.812	0.838	0.864	0.891
0.5	0.721	0.757	0.793	0.830	0.866	0.902	0.938	0.974	1.010	1.046
0.6	0.783	0.830	0.877	0.924	0.971	1.018	1.065	1.112	1.159	1.206
0.7	0.841	0.900	0.959	1.018	1.077	1.135	1.194	1.253	1.312	1.371
0.8	0.896	0.968	1.040	1.111	1.183	1.255	1.326	1.398	1.470	1.542
0.9	0.949	1.035	1.120	1.205	1.291	1.376	1.462	1.547	1.632	1.718

poor reflectors at wavelengths below 340 nm ($R = 0.30$ at 250 nm, for example). Experimenters who have access to vacuum evaporation equipment of the type used for shadowing electron microscope specimens can readily make efficient reflectors by condensing evaporated aluminum on the surfaces of reaction vessels. If the vessel is made of glass it is essential to make the coating on the inside surface, since the high absorptivity of glass in the ultraviolet region would nullify any advantage conferred by a reflecting surface behind it. Aluminum films in contact with aqueous solutions are not very stable, however, and have the added disadvantage of a lethal action (Yamamoto *et al.*, 1964) against certain viruses.

Second surface reflectors prepared by depositing aluminum on the bottom surfaces of quartz cavity slides (see Fig. 4) have reflectance coefficients as high as 0.85. For similar vessels machined from aluminum alloy and given a high polish, $R = 0.6$; and for chromium-plated brass, $R = 0.4$ at 254 nm.

For virus suspensions which are perfectly mixed during t seconds of irradiation at average intensity $\bar{I}$, the quantum dose of light energy is $\bar{I}t$, and the survival ratio for a first-order inactivation process is

$$\frac{N}{N_0} = e^{-k\bar{I}t} \tag{3}$$

in which N is the number of infective units of virus surviving at time t, N_0 is the number originally present, and k is a velocity constant characteristic of the particular virus and the wavelength of ultraviolet light employed. Note that the units of k are such that the product $k\bar{I}t$ is unitless. If t is in seconds and $\bar{I}$ is in ergs mm^{-2} sec^{-1}, then the units of k are mm^{2} erg^{-1}.

G. Effect of Stratification

The opposite to the condition of perfect mixing is no mixing at all. This condition is approached closely when very short exposure times are used. The volume of fluid being irradiated may be regarded as a stack of infinitely thin strata, each at a different distance (x) from the plane of the incident light. As Morowitz (1950) pointed out, the light intensity in each stratum is $I_0e^{-\alpha x}$ and the survival ratio is

$$\frac{n}{n_0} = e^{-kI_0e^{-\alpha x}t} \tag{4}$$

When the suspension is mixed after irradiation, the survival ratio in the total volume is

$$\frac{N}{N_0} = \frac{1}{L}\int_0^L e^{-kI_0e^{-\alpha x}t}\, dx \tag{5}$$

When the additional complication of a reflective supporting surface is introduced (Hiatt, 1961), Eq. (5) becomes

$$\frac{N}{N_0} = \frac{1}{L}\int_0^L e^{kI_0t}(e^{-\alpha x} + Re^{-2\alpha L}e^{\alpha x})\, dx \tag{6}$$

Computer evaluations disclose that Eq. (6) converges with Eq. (3) for values of $e^{-\alpha L}$ close to 1. Thus when the liquid layer is thin enough or transparent enough, it makes no difference whether it is mixed or not.

Table IV lists the limiting values of transmittance ($e^{-\alpha L}$) above which

TABLE IV

Values of Transmittance ($e^{-\alpha L}$) above Which Mixing of the Liquid during Irradiation Does Not Appreciably Affect the Predicted Survival Ratio[a]

Reflection coefficient (R)	Transmittance ($e^{-\alpha L}$)
0	0.84
0.1	0.81
0.2	0.78
0.3	0.75
0.4	0.72
0.5	0.69
0.6	0.66
0.7	0.63
0.8	0.60
0.9	0.57

[a] Data from Hiatt (1961).

the difference between perfect mixing and complete stratification amounts to less than 0.1 logarithmic unit in the survival ratio. In view of the usual degree of uncertainty in microbiological assays, it appears that, by working with liquid layers which transmit 85% or more of the incident light, the experimenter may avoid the necessity of mixing the liquid during irradiation. (For liquids containing cellular debris or very large particles, stirring may be required, as earlier noted, simply to maintain a uniform suspension.) Uniform mixing of small volumes of liquid is difficult to achieve, and there is a distinct advantage to working under conditions which make the efficiency of mixing less than crucially important.

H. Interpretation of Survival Curves

Ultraviolet inactivation of many viruses proceeds as a first-order function of quantum dose, as represented in Eq. (3). In logarithmic form, the equation becomes

$$\log \frac{N}{N_0} = -\frac{k\bar{I}t}{2.303} \tag{7}$$

and when log N/N_0 is plotted against quantum dose, as in Fig. 7, the result is a straight line. (In the example cited, time of irradiation at constant intensity is taken as a measure of the quantum dose and the velocity constant for the reaction then becomes $k\bar{I}$.)

Baron *et al.* (1959) reported the results shown in Fig. 8 for a variety of animal viruses irradiated at 254 nm with a constant incident intensity

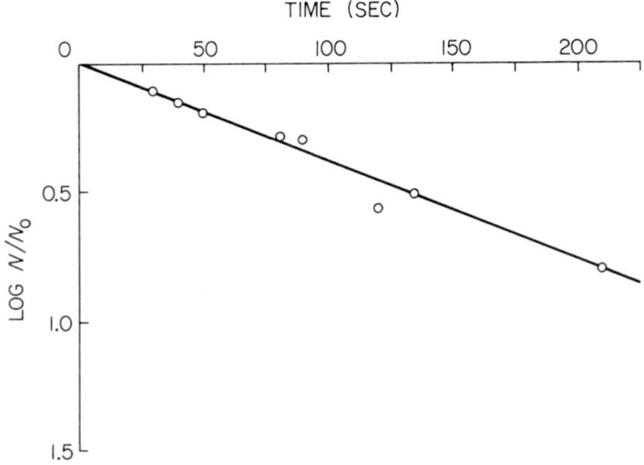

Fig. 7. Semilogarithmic survival curve for tobacco mosaic virus as a function of time of irradiation at 253.7 nm. (Data obtained from Oster and McLaren, 1950.)

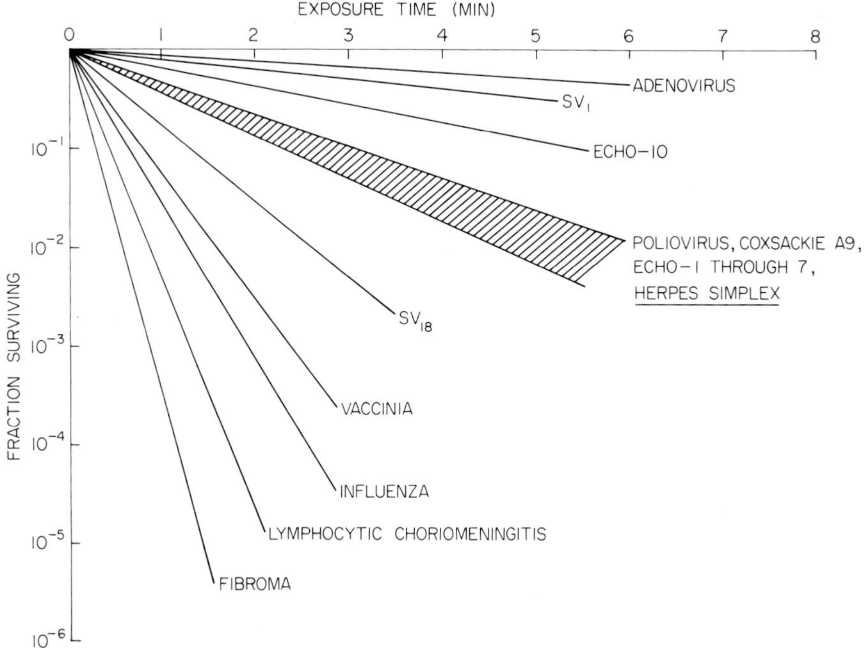

Fig. 8. Comparative ultraviolet sensitivities of a variety of animal viruses. [Data from Baron *et al.* (1959) reprinted from Hiatt (1962).]

of 18 $\mu W/cm^2$. The differences in reaction rate appeared to correlate with accepted taxonomic groupings of the viral agents.

Survival curves which, when plotted in semilogarithmic form, have a distinct lag phase or initial shoulder are usually characterized by extrapolating back from the linear portion of the curve to obtain a value for the y-intercept, called the "extrapolation number." According to target theory, a virus which has n replicate sites of reaction, all of which must be destroyed to inactivate the particle, would follow the survival equation

$$\frac{N}{N_0} = 1 - (1 - e^{-k\bar{I}t})^n \tag{8}$$

The shape of this function in logarithmic form is shown (for $n = 4$) in curve C, Fig. 9. Curve B is the extrapolation line, drawn back to the y-axis where it intercepts at $\log 4 = 0.602$. To see how this intercept relates to n in Eq. (8), you can expand the polynomial to obtain

$$\frac{N}{N_0} = 1 - \left(1 - ne^{-k\bar{I}t} + \frac{n(n-1)e^{-2k\bar{I}t}}{2!} + \cdots\right) \tag{9}$$

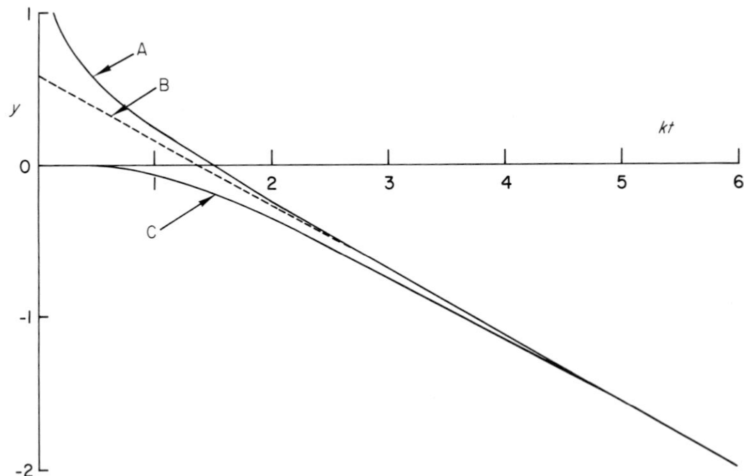

Fig. 9. Theoretical survival curves for a virus which has four essential targets. (A) The Atwood–Norman transformation, $y = \log\{-\ln(1 - N/N_0)\}$; (B) the extrapolation line, $y = \log(ne^{-kt})$; and (C) the survival function as predicted by target theory, $y = \log N/N_0 = \log\{1 - (1 - e^{-kt})^4\}$. See text for discussion.

the second and subsequent terms of the series become negligible for high values of $k\bar{I}t$, and under these conditions

$$\frac{N}{N_0} \cong ne^{-k\bar{I}t} \tag{10}$$

The value of the intercept is obtained by setting $k\bar{I}t = 0$, for which $N/N_0 = n$.

The extrapolation line is often difficult to draw objectively, and experimental values for n are sometimes considerably in error. Atwood and Norman (1949) suggested a transformation to improve the accuracy of this extrapolation. Instead of log N/N_0, you plot log $\{-\ln[1 - (N/N_0)]\}$ versus dose (see curve A in Fig. 9). The best straight line fitted by least squares gives a more dependable value for n because it makes use of the data for high survival ratios.

Another type of survival curve sometimes encountered in ultraviolet irradiation experiments with viruses is seen in Fig. 10. Here the positive curvature in the semilogarithmic plot indicates that the inactivating process becomes less effective as the dose increases, or that there are some virions which are more resistant than others. The implications of this type of survival curve have been discussed by Hiatt (1964). The two most probable explanations for the curvature are (1) the virus preparation is initially heterogenous with respect to ultraviolet sensitivity, or (2) an

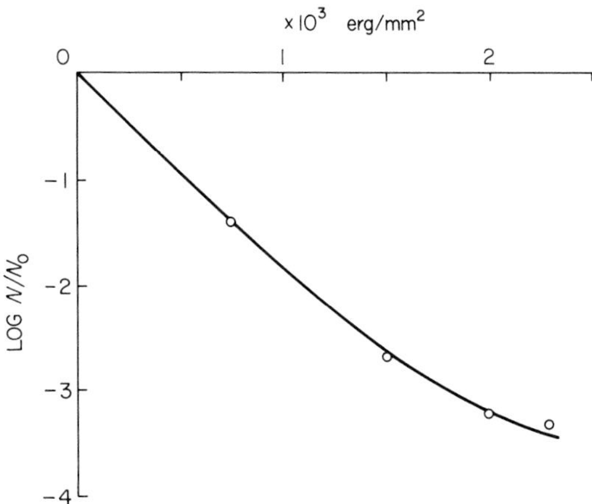

Fig. 10. Survival of poliovirus (type 2) as a function of quantum dose at 253.7 nm. [Data obtained in collaboration with George D. Gardner, National Institutes of Health (1957).]

initially homogeneous preparation develops heterogeneity in the course of the inactivating process.

The mathematical expression for survival of an initially heterogenous virus suspension is

$$\frac{N}{N_0} = \gamma_1 e^{-k_1 \bar{I} t} + \gamma_2 e^{-k_2 \bar{I} t} + \gamma_3 e^{-k_3 \bar{I} t} + \cdots \tag{11}$$

in which γ_1 is the fraction of virus which is inactivated with velocity constant k_1, γ_2 the fraction with velocity constant k_2, etc. If there are only two types of virus initially present: a normal type, with velocity k_1 and a resistant type, with velocity k_2, then

$$\frac{N}{N_0} = (1 - \gamma) e^{-k_1 \bar{I} t} + \gamma e^{-k_2 \bar{I} t} \tag{12}$$

in which γ is the resistant fraction. Since $k_1 \gg k_2$, at high values of $\bar{I}t$ the survival ratio is approximately

$$\frac{N}{N_0} \cong \gamma e^{-k_2 \bar{I} t} \tag{13}$$

and extrapolating back to the y-axis (setting $\bar{I}t = 0$) gives an intercept which is equal to γ.

If the virus preparation is initially homogeneous, but develops heterogeneity during irradiation, the same curve shape may result. The inactivation reaction, in general form is

$$
\begin{array}{ccc}
 & & B \xrightarrow{k_2} X \\
 & k_1 \nearrow & \\
A & & \\
 & k_3 \searrow & \\
 & & X
\end{array}
$$

in which A represents the virus in its initial state and B represents a damaged virion which is still viable, but greatly reduced in efficiency. Dead virus is represented by X. If the efficiency of the normal virus is 1 and that of the damaged virus is γ, then the survival ratio may be expressed as

$$\frac{N}{N_0} = \frac{A + \gamma B}{A_0} \tag{14}$$

in which A_0 is the number of normal virions originally present and A and B represent the number of normal and damaged virions, respectively, that are present at time t.

The mathematical model for this system consists of three differential equations:

$$\frac{dA}{dt} = -(k_1 + k_3)A \tag{15}$$

$$\frac{dB}{dt} = k_1 A - k_2 B \tag{16}$$

$$\frac{dX}{dt} = k_3 A + k_2 B \tag{17}$$

and the conservation equation:

$$A + B + X = A_0 \tag{18}$$

If, for simplicity, we make $k_3 = k_2$, the solution of the equation set gives

$$A = A_0 e^{-(k_1+k_2)t} \tag{19}$$

$$B = A_0(e^{-k_2 t} - e^{-(k_1+k_2)t}) \tag{20}$$

By substitution into Eq. (14) we obtain

$$\frac{N}{N_0} = (1 - \gamma)e^{-(k_1+k_2)t} + \gamma e^{-k_2 t} \tag{21}$$

which is virtually identical with Eq. (12).

Of the two explanations considered, the latter seems more plausible since many viruses have been found to depend upon their protein coats for invasiveness but to remain infective (at much lower efficiency) when their coats are removed or damaged.

Among the other complications which may affect the shape of survival curves are multiplicity reactivation (Luria, 1955), viral aggregation (Sharp and Kim, 1966), and several types of repair or reversal of ultraviolet damage (cf. Smith and Hanawalt, 1969).

I. Continuous-Flow Irradiation

Following the work of Hodes *et al.* (1940), which demonstrated the feasibility of using ultraviolet light to inactivate rabies virus for vaccine production, several groups of investigators became interested in developing more efficient means of irradiating large volumes of liquid. Oppenheimer and Levinson (1943) designed the first of a series of centrifugal filmers which maintained a continuously flowing thin film of liquid on the inside surface of a hollow rotor which surrounded a low-pressure mercury arc lamp. Although the description of this device was withheld from general publication because of military restrictions, it was used successfully by Levinson *et al.* (1944, 1945) for the production of experimental vaccines.

McLean *et al.* (1945) independently devised a centrifugal filming device which consisted of a horizontal rotating cylinder with an axial tubular lamp. This device was used to produce experimental influenza vaccines. Habel and Sockrider (1947) developed a cylindrical filming device which was inclined at a slight angle so that fluids would move through it by gravity flow. Bozeman *et al.* (1950) constructed a modified version of the Habel–Sockrider device and developed monitor systems to assure continuity of fluid flow and illumination. Variations of this device are still in use in the biologics industry for sterilization of injectable fluids and for inactivation of viral vaccines.

The original Oppenheimer–Levinson apparatus was improved in design (Oppenheimer *et al.*, 1959) and was used commercially as a supplementary inactivation procedure for Salk poliomyelitis vaccine. The construction of the centrifugal filmer is shown schematically in Fig. 11. A stainless steel bowl (A) is rotated at 1750 rpm around a vertical axis. The inside surface of the bowl is conical, with a 2° taper, and the liquid to be irradiated, when introduced at the bottom, flows upward as a thin film stabilized in its vertical dimension by increasing centrifugal acceleration as the liqud ascends. After reaching the top of the bowl, the liquid is propelled into the collecting cover (C1) and delivered to the discharge outlet (D). A bank of six tubular germicidal lamps is arranged symmetrically around the axis of rotation, and the lampholder (B) is ported to allow introduction of sterile inert gas during irradiation.

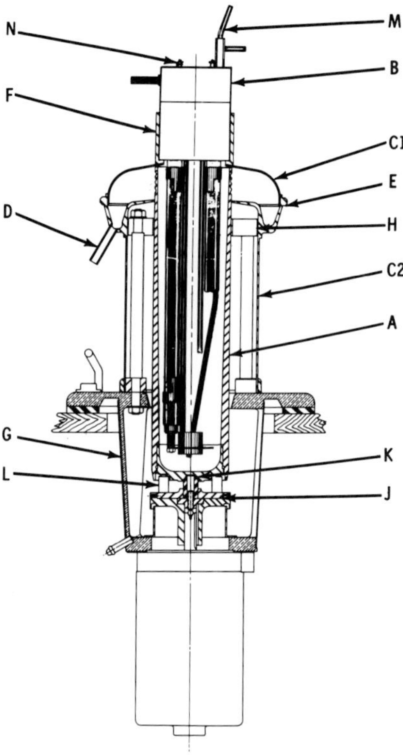

Fig. 11. Schematic drawing of a modern centrifugal film irradiation device. [Reprinted from Oppenheimer *et al.* (1959) with permission from the American Public Health Association.]

When the centrifugal filmer is used for continuous-flow sterilization, the configuration of equipment is as depicted in Fig. 12. Aside from the obvious need to maintain a microbiologically closed system when handling the flowing liquid, this method of sterilization demands stringent process controls. With a film thickness of 100 μ or less, which must be maintained within close limits to assure uniform light penetration, the irradiated fluid has a dwell time of about 1 sec. If for any reason the fluid traversing the bowl during this brief period fails to receive its full dose of light energy, viable organisms may pass through and contaminate the entire lot of processed material. Recording monitors with response times demonstrated to be adequate must be used to verify that there are no interruptions in illumination or discontinuities in flow. In spite of the formidable requirements it imposes and some disappointment over the fact that it is not universally effective, ultraviolet sterilization has been a useful proce-

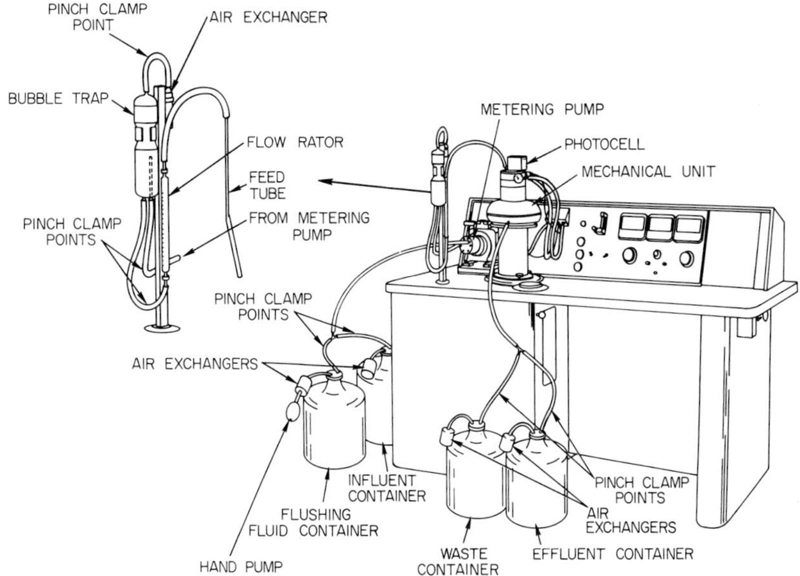

Fig. 12. A commercial model of the Oppenheimer–Benesi–Taylor centrifilmer, showing the array of ancillary equipment required for continuous-flow sterilization of liquids. (Courtesy of Beckman Instruments, Inc., Spinco Division, Palo Alto, Calif.)

dure in the biologics industry and will probably continue to be used for processes in which its capability has been demonstrated.

III. Photodynamic Action

A. Review

Photodynamic action is the name given to the class of photochemical reactions that destroy the structure or function of a biological substrate through the intermediate action of a photosensitizer. The photosensitizer acts as a catalyst and is not consumed in the reaction. Oxygen is required as a reactant in some, but not all, photodynamic processes. Many of the photosensitizers which have been studied are organic dyes with absorption peaks in the visible range. Light of wavelength between 320 and 700 nm satisfies the quantum energy requirements of the reaction.

Blum (1941) wrote a comprehensive monograph on photodynamic action which has served as a basic reference for a generation of investigators. Other reviews by Fowlks (1959), Santamaria (1960), McLaren and Shugar (1964), and Spikes (1968) summarize the broad aspects of the subject.

Destruction of the infectivity of viruses by photodynamic action was observed by Herzberg (1933), Perdrau and Todd (1933a), and Shortt and Brooks (1934), among others, and the possibility of using photodynamic action to prepare viral vaccines was investigated by Perdrau and Todd (1933b), Dempsey and Mayer (1934), and Galloway (1934). Although the results reported were quite encouraging, no practical application developed and the subject lay dormant for 20 years. Heinmets *et al.* (1955) investigated photodynamic action as one of several possible methods for inactivating the virus of serum hepatitis in human plasma.

Welsh and Adams (1954) explored the use of photodynamic inactivation rate as an aid to the classification of bacteriophages, extending the earlier work of Burnet (1933). Yamamoto (1958) also found characteristic differences in photodynamic inactivation rates among bacteriophages, with an especially pronounced difference between T-odd and T-even coliphages. Helprin and Hiatt (1959) showed that this difference resulted from a disparity in the rate of uptake of photosensitizer.

Hiatt *et al.* (1960) tested a variety of animal viruses for photodynamic sensitivity and found that poliovirus was inherently resistant to photosensitization. This observation led to the development of a set of procedures for differential inactivation of contaminating viruses in live attenuated poliovirus vaccine (cf. Hiatt, 1960a,b, 1962).

The observations of Crowther and Melnick (1961) and Schaffer (1962), disclosing that poliovirus could be photosensitized if the sensitizer were present during intracellular maturation of the virus, led to the exploitation of this property to obtain information about the early events in viral synthesis (Wilson and Cooper, 1962, 1963; Schaffer and Hackett, 1963; Hiatt and Moore, 1965) and to provide a means of tracing the virus by a lethal tag (Hiatt and Moore, 1962, 1963; Cords and Holland, 1964; Miller and Horstmann, 1966).

The photodynamic inactivation rates of many animal viruses have now been reported (Moore *et al.*, 1962; Hiatt *et al.*, 1960, Orlob, 1962, 1963; Tomita and Prince, 1963; Scholtissek and Rott, 1964; Thormar and Petersen, 1964; Wallis and Melnick, 1964; Sinkovics *et al.*, 1965). Wallis and Melnick (1963) found that the characteristic photodynamic resistance of poliovirus vanished if the virus was physically purified and resuspended in an alkaline milieu. Experimental influenza vaccines prepared by photodynamic action were found by Wallis *et al.* (1963) to be comparable in potency to formalin-inactivated vaccines.

B. Photosensitizers

The compounds found to be photochemically active against viruses for the most part are heterotricyclic dyes of the phenazine, thiazine, or acridine series. Figure 13 shows the structural formulas and visible light absorption spectra of representative compounds in each of these series. The cytotoxicity of each of these compounds is relatively low, permitting their use at concentrations of 2 to 6 μg/ml in many tissue culture systems.

The choice of dye concentration for photodynamic action experiments is necessarily a compromise between two considerations. The higher the dye concentration, the higher will be the reaction rate for a given average light intensity. But the effect of optical shielding is in the opposite direction, and if too much dye is used, the reaction may be drastically retarded. The rate of change of virus concentration with time, in a perfectly mixed volume of fluid is

$$\frac{dN}{dt} = -kc\bar{I}N \tag{22}$$

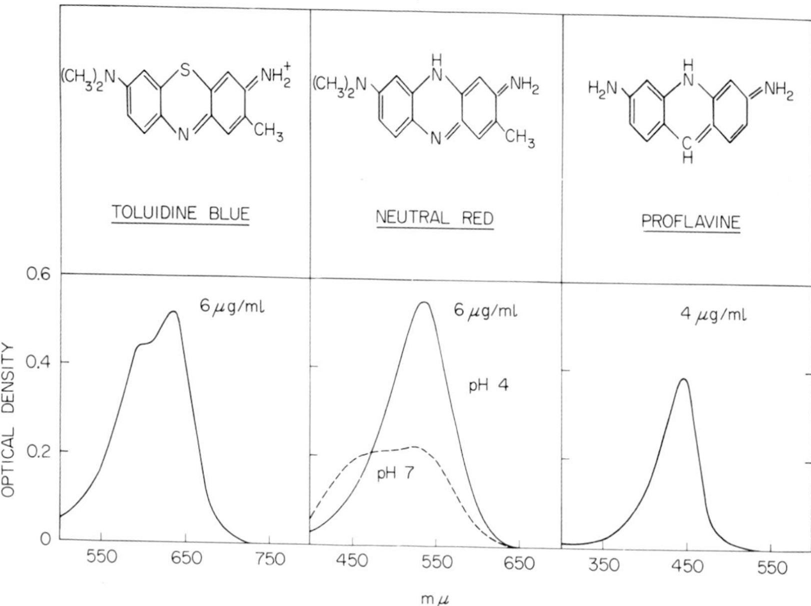

Fig. 13. Structural formulas and visible light absorption spectra for three photodynamically active dyes. [From Hiatt (1960b), reprinted with permission of the New York Academy of Sciences.]

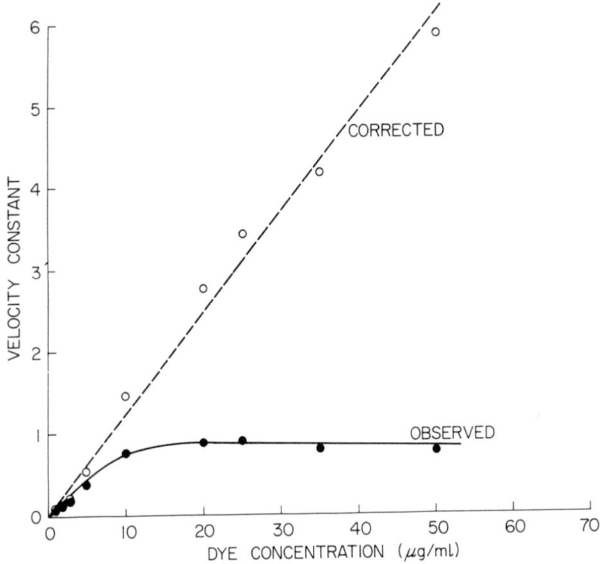

Fig. 14. Effect of dye concentration upon the velocity constant for photodynamic inactivation of T1 coliphage with methylene blue. Solid circles are data from Welsh and Adams (1954); open circles are the same data corrected for optical shielding as described in the text. [Figure reprinted from Hiatt (1960b) with permission of the New York Academy of Sciences.]

for a single target reaction in which c is the concentration of photosensitizer in the vicinity of the reaction sites.

Figure 14 shows the effect of optical shielding upon the observed reaction velocity for photodynamic inactivation of phage. The corrected curve illustrates the theoretical effect of increasing dye concentration if there were no optical shielding.

C. Reaction Mechanisms

The first step in the photodynamic inactivation of a virus is the combination of the photosensitizer D with the critical reaction sites of the virus particle, represented as V, and this reaction may be written as

$$D + V \leftrightharpoons D \cdot V$$

in which D·V is a dye–virus complex in the ground state. As originally noted by Oster and McLaren (1950) in experiments with acridine orange and tobacco mosaic virus, it is necessary that the dye be combined with the virus before the photochemical steps in the reaction can begin. The combination of dye with some viruses is very slow. The slow permeation

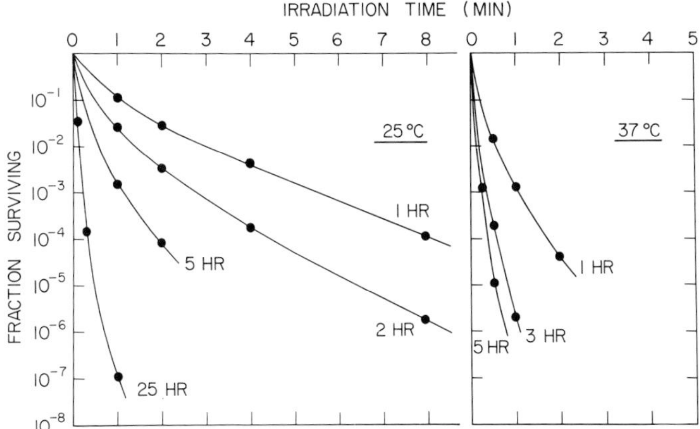

Fig. 15. Relative velocity constant for the inactivation of T2 coliphage as a function of the length of incubation of the dye–phage mixtures at 25° or 37°C prior to irradiation. (From Helprin and Hiatt, 1959.)

of toluidine blue into the vital areas of T2 coliphage, for example, may take days or weeks at 4°C to reach a state of equilibrium (Hiatt, 1960b) and even at 25° or 37°C is incomplete after many hours (see Fig. 15). For other viruses, typically the T-odd coliphages (Fig. 16), the reaction is

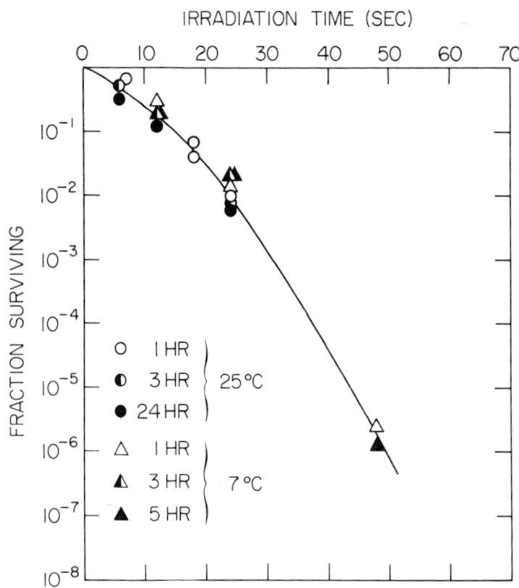

Fig. 16. Survival curves for T3 coliphage irradiated with polychromatic light ($I_0 = 260\ \mu W/cm^2$) after dye–phage mixtures were incubated for the times and at the temperatures indicated before irradiation. (From Helprin and Hiatt, 1959.)

quite rapid, equilibrium being reached within a second or less after mixing.

After formation of the dye–virus complex, the photochemical steps of the reaction are initiated by absorption of a photon ($h\nu$) of the proper quantum energy

$$D \cdot V + h\nu \rightarrow (D \cdot V)^*$$

to yield an excited state of some portion of the complex $(D \cdot V)^*$.

Through a series of energy transfers, as yet controversial in nature, the excited complex reacts in such a way (sometimes requiring O_2) as to destroy the replicating ability of the virus

$$(D \cdot V)^* \xrightarrow{O_2} D \cdot V'$$

and the inactive virus (V') is in equilibrium with unaltered dye,

$$D \cdot V' \leftrightarrows D + V'$$

The chemical lesions that kill the virus are not known in all cases, but the findings of Simon and Van Vunakis (1962), Sastry and Gordon (1966), Waskell *et al.* (1966), and Freifelder and Uretz (1966) point to preferential destruction of guanine residues in the viral DNA as the most probable lethal reaction when methylene blue or acridine orange are used as photosensitizers. Guanine is also a preferred reaction point in the RNA of tobacco mosaic virus, according to Singer and Fraenkel-Conrat (1966), but the number of guanine residues affected greatly exceeds the number of lethal hits, indicating that not all guanine residues are important to virus survival. Since the critical reaction sites for many viruses are evidently part of the viral nucleic acid, as predicted earlier by Yamamoto (1958), and the protein capsids, which contain the viral antigens, are comparatively resistant to photodynamic action, the possibility of making highly potent killed-virus vaccines by this method is attractive from a theoretical point of view.

D. Kinetics

Most of the viruses inactivated by photodynamic action behave as single targets, following a first-order decay process with respect to quantum dose. One of the exceptions is T3 coliphage sensitized with toluidine blue (see Fig. 16). The initial shoulder of this curve, which persists over a wide range of experimental conditions (Hiatt, 1966), has an extrapolation number of approximately 4, and seems truly to represent cumulative damage to the virion. At high light intensities, semilogarithmic survival curves for the T-even coliphages sensitized with toluidine blue are dis-

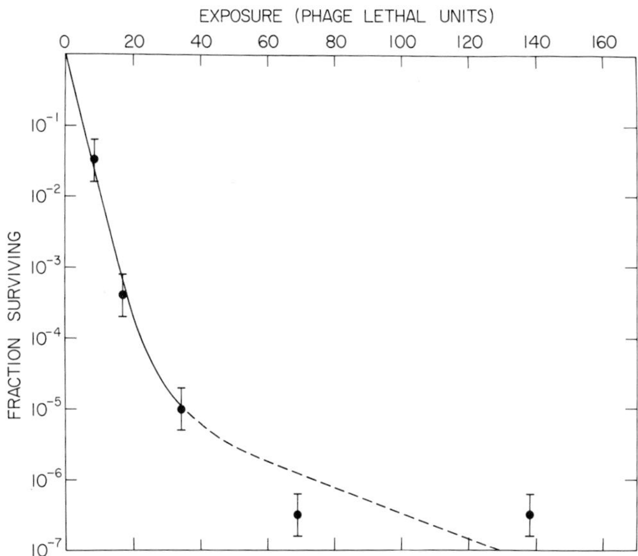

Fig. 17. Survival curve for SV40 irradiated with polychromatic light in the presence of toluidine blue, 6 μg/ml. The units of quantum dose are phage lethal units (PLU). One PLU is the dose required to reduce the survival ratio of a T3 phage suspension to e^{-1}. (From Hiatt *et al.,* 1962.)

tinctly curvilinear, as shown by the family of curves in Fig. 15. At lower light intensities the semilogarithmic plots are nearly straight lines. Oxygen depletion within the phage particle has been offered as possible explanation for this effect (Hiatt, 1966).

Survival curves which are biphasic, or which have "tailing effects" of unexplained origin, are sometimes seen in photodynamic inactivation experiments (as shown, for example, in Fig. 17), and impose some limits upon the usefulness of photodynamic action as a tool for making viral vaccines. All inactivating processes that have been used for the commercial processing of viral vaccines have some similar limitations, however, and in this context photodynamic action remains as a method of possible future interest.

E. Apparatus

In the visible light range (320–700 nm), clarified virus suspensions and the glass vessels in which they are ordinarily contained are virtually transparent. The absorbency of the added photosensitizer and its combination products therefore determines the degree of optical shielding that can occur during irradiation. Photodynamic action experiments, in contrast to UV reactions, may be conducted with relatively thick layers of liquid and with the convenience of using closed glass containers.

Preliminary experiments have often been carried out by simply exposing test tubes, vials, bottles, or petri dishes containing the test liquids to a standard incandescent or fluorescent light source for graded intervals of time. For experiments in which stirring is desirable, the apparatus shown in Fig. 18 is convenient to use. The disposable aluminum weighing dishes containing the liquid samples have favorable optical properties (R = ca. 0.9), but in some sensitive systems the aluminum surface may have a virucidal effect (Yamamoto *et al.*, 1964).

Polychromatic light sources are appropriate for preliminary experiments because they are inexpensive, readily available, and have abundant energy. Incandescent tungsten filament lamps are probably the most versatile of the sources because of the continuity and relative flatness of their emission spectrum (see Fig. 19). The high-temperature tungsten–

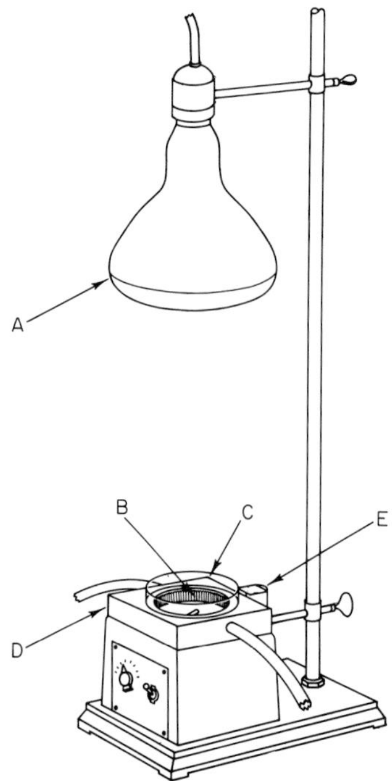

Fig. 18. Device for exposing virus specimens to polychromatic light. (A) General Electric RFL-2 photoflood lamp, (B) disposable aluminum-foil dish, (C) petri dish cover, (D) water-jacketed receptacle, (E) photoelectric cell (International Rectifier B2M). (Reprinted from Hiatt *et al.*, 1960.)

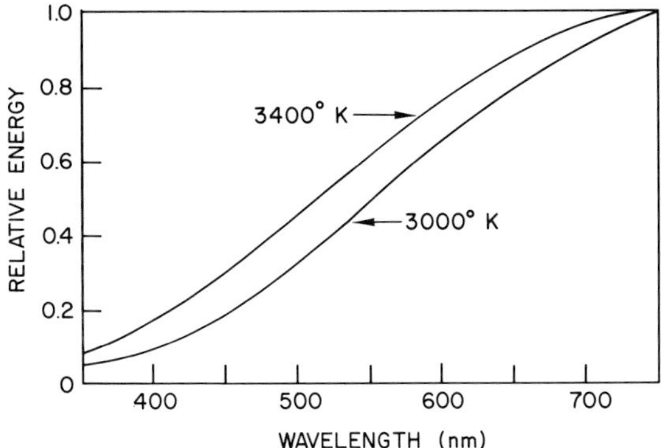

Fig. 19. Spectral energy distribution of conventional tungsten or tungsten–halide lamps at color temperatures of 3000° and 3400°K. (Adapted from Lighting Handbook, 3rd ed., 1969, Sylvania Electric Products, Inc., Danvers, Mass.)

halide lamps are ideally compact and powerful sources with emission spectra simulating that of the sun. (A clear glass filter should be used with these sources to cut off light of wavelength less than 290 nm.) Fluorescent lamps of various sizes have the advantages afforded by their cool operating temperature, but they are relatively deficient in red light

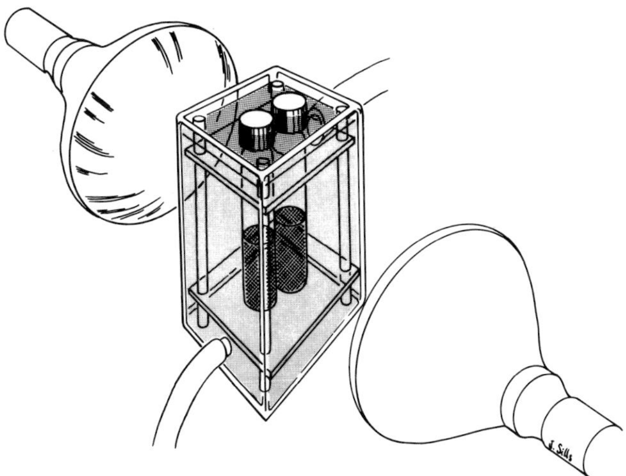

Fig. 20. Simple device for illuminating virus suspensions in closed vessels. Two General Electric RFL-2 photoflood lamps at a face-to-face distance of 15 cm are on either side of a Pyrex aquarium jar converted to a flowing-water bath. A Plexiglas rack holds two or more reaction tubes. [Reprinted from Hiatt (1960b) with permission of the New York Academy of Sciences.]

and thus are not ideal for use with the many blue dyes which are active photosensitizers.

Experiments with monochromatic light require a monochromator or appropriate interference filter. The emergent light, in either case, is a partially collimated beam of light. The apparatus depicted in Fig. 4 is especially useful for this type of experiment, since the coplanar thermopile may be positioned in the light beam to measure incident irradiance.

With highly infectious or dangerous pathogenic viruses, experiments with polychromatic light may be conducted with apparatus as shown in Fig. 20, temperature control being achieved by rapid circulation of cold tap water.

Continuous-flow irradiation of biological fluids with visible light is comparatively simple in its apparatus requirements because of the practicability of using glass tubing as a conduit for the fluid and cold water as a heat sink. In 1955 Heinmets *et al.* described the apparatus shown in Fig. 21 as a device of possible application to the sterilization of human plasma. With toluidine blue (10^{-5} M) as photosensitizer, human plasma could be freed of added eastern equine encephalomyelitis virus by passing it through the irradiator at flow rates as high as 200 ml/min. A refinement of this device, utilizing a compact cylindrical source of light (Fig. 22), was used experimentally for differential inactivation of adventitious viruses in live attenuated poliovirus vaccine (Hiatt, 1960a, 1962).

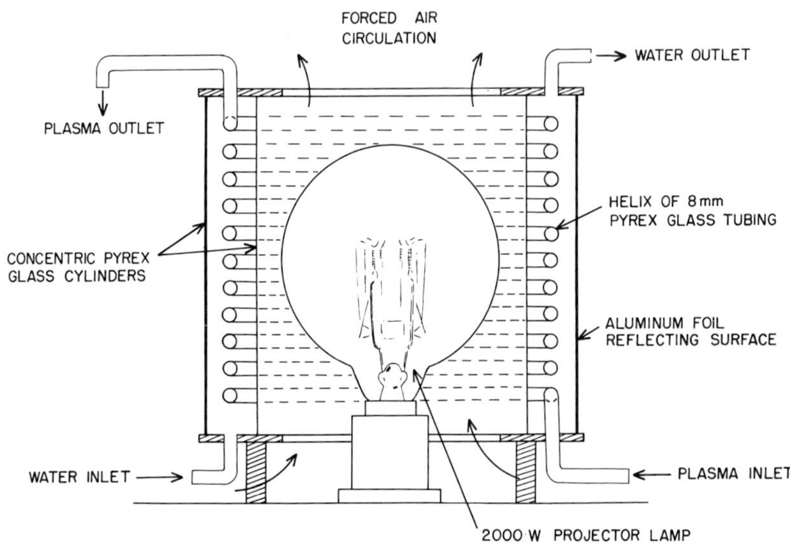

Fig. 21. Continuous-flow irradiator for processing human plasma. (Reprinted from Heinmets *et al.*, 1955.)

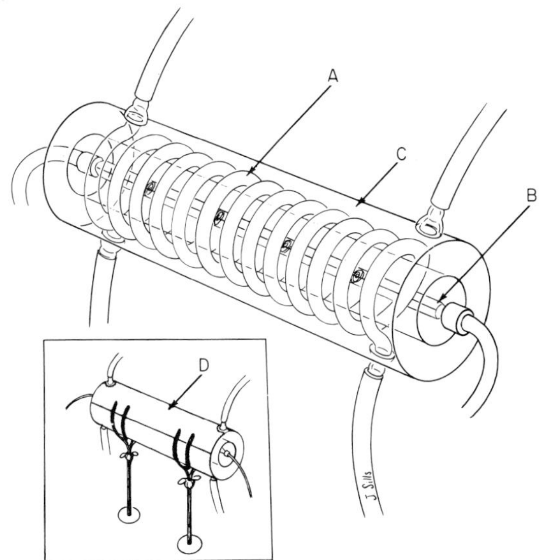

Fig. 22. Apparatus for continuous-flow irradiation of liquids. The liquid to be treated is pumped through a helical coil of Pyrex glass tubing (A) around the light source (B), which is a General Electric Quartzline 1500T3Q/CL tungsten filament lamp. The coil is surrounded by a water jacket (C) and a polished aluminum reflector (D). [Reprinted from Hiatt (1960b) with permission of the New York Academy of Sciences.]

Although there has been no sustained commercial use of continuous-flow photodynamic inactivation processes in biologics production, both the simplicity of the required apparatus and the inherent selectivity of action of these processes commend them for further consideration.

References

Atwood, K. C., and Norman, A. (1949). *Proc. Nat. Acad. Sci. U. S.* **35,** 696.

Baron, S., Miller, A. H., Gochenour, A. M., and Hiatt, C. W. (1959). *Fed. Proc., Fed. Amer. Soc. Exp. Biol.* **18,** 557.

Bawden, F. C., and Kleczkowski, A. (1953). *J. Gen. Microbiol.* **8,** 145.

Blum, H. F. (1941). "Photodynamic Action and Diseases Caused by Light." Von-Nostrand-Reinhold, Princeton, New Jersey (republished by Hafner, New York, 1964).

Bowen, E. J. (1946). "The Chemical Aspects of Light," 2nd ed., p. 283. Oxford Univ. Press (Clarendon), London and New York.

Bozeman, V., Tripp, J. T., and Berry, B. (1950). *J. Immunol.* **64,** 65.

Burnet, F. M. (1933). *J. Pathol. Bacteriol.* **37,** 179.

Chessin, M. (1960). *Science* **132,** 1840.

Cords, C. F., and Holland, J. J. (1964). *Proc. Nat. Acad. Sci. U. S.* **51,** 1080.

Crowther, D., and Melnick, J. L. (1961). *Virology* **14,** 11.
Dempsey, T. F., and Mayer, V. (1934). *J. Comp. Path. Therap.* **47,** 197.
Fluke, D. J., and Pollard, E. C. (1949). *Science* **110,** 274.
Fowlks, W. L. (1959). *J. Invest. Dermatol.* **32,** 233.
Freifelder, D., and Uretz, R. B. (1966). *Virology* **30,** 97.
Galloway, I. A. (1934). *Brit. J. Exp. Pathol.* **15,** 97.
Habel, K., and Sockrider, B. T. (1947). *J. Immunol.* **56,** 273.
Hass, G. (1955). *J. Opt. Soc. Amer.* **45,** 945.
Hatchard, C. G., and Parker, C. A. (1956). *Proc. Roy. Soc., Ser. A* **235,** 518.
Heinmets, R., Kingston, J. R., and Hiatt, C. W. (1955). Rep. No. WRAIR-53-55. Walter Reed Army Inst. Res., Washington, D. C.
Helprin, J. J., and Hiatt, C. W. (1959). *J. Bacteriol.* **77,** 502.
Herzberg, K. (1933). *Z. Immunitaetsforschu. Exp. Ther.* **80,** 507.
Hiatt, C. W. (1960a). *Fed. Proc., Fed. Amer. Soc. Exp. Biol.* **19,** 405.
Hiatt, C. W. (1960b). *Trans. N. Y. Acad. Sci.* [2] **23,** 66.
Hiatt, C. W. (1961). *Nature (London)* **189,** 678.
Hiatt, C. W. (1962). *Proc. Int. Congr. Microbiol. Stand., 7th, 1961* p. 118.
Hiatt, C. W. (1964). *Bacteriol. Rev.* **28,** 150.
Hiatt, C. W. (1966). *In* "Radiation Research" (G. Silini, ed.), p. 857. Wiley, New York.
Hiatt, C. W., and Moore, D. (1962). *Fed. Proc., Fed. Amer. Soc. Exp. Biol.* **21,** 460.
Hiatt, C. W., and Moore, D. (1963). *Fed. Proc., Fed. Amer. Soc. Exp. Biol.* **22,** 558.
Hiatt, C. W., and Moore, D. E. (1965). *Proc. Soc. Exp. Biol. Med.* **119,** 203.
Hiatt, C. W., Gerber, P., and Friedman, R. M. (1962). *Proc. Soc. Exp. Biol. Med.* **109,** 230.
Hiatt, C. W., Kaufman, E., Helprin, J. J., and Baron, S. (1960). *J. Immunol.* **84,** 480.
Hodes, H. L., Webster, L. T., and Lavin, G. I. (1940). *J. Exp. Med.* **72,** 437.
Jagger, J. (1961). *Radiat. Res.* **14,** 394.
Jagger, J. (1967). "Introduction to Research in Ultraviolet Photobiology." Prentice-Hall, Englewood Cliffs, New Jersey.
Koller, L. R. (1939). *J. Appl. Phys.* **10,** 624.
Latarjet, R., and Wahl, R. (1945). *Ann. Inst. Pasteur, Paris* **71,** 336.
Latarjet, R., Morenne, P., and Berger, R. (1953). *Ann. Inst. Pasteur, Paris* **85,** 174.
Levinson, S. O., Milzer, A., Shaughnessy, H. J., Neal, J. L., and Oppenheimer, F. (1944). *J. Amer. Med. Ass.* **125,** 531.
Levinson, S. O., Milzer, A., Shaughnessy, H. J., Neal, J. L., and Oppenheimer, F. (1945). *J. Immunol.* **50,** 317.
Luria, S. E. (1955). *Radiat. Biol.* **2,** 333.
McLaren, A. D., and Moring-Claesson, I. (1961). *Progr. Photobiol., Proc. Int. Congr., 3rd, 1960* pp. 573–575.
McLaren, A. D., and Shugar, D. (1964). "Photochemistry of Proteins and Nucleic Acids." Macmillan, New York.
McLaren, A. D., and Takahashi, W. N. (1957). *Radiat. Res.* **6,** 532.
McLean, I. W., Jr., Beard, D., Taylor, A. R., Sharp, D. G., and Beard, J. W. (1945). *J. Immunol.* **51,** 65.
Miller, D. G., and Horstmann, D. M. (1966). *Virology* **30,** 319.
Moore, D., Helprin, J. J., and Hiatt, C. W. (1962). *Bacteriol. Proc.* p. 150.
Morowitz, H. J. (1950). *Science* **111,** 229.
Oppenheimer, F., and Levinson, S. O. (1943). Publication withheld by the Committee on Medical Research of the Office of Scientific Research and Development, U. S.

Oppenheimer, F., Benesi, E., and Taylor, A. R. (1959). *Amer. J. Pub. Health Nat. Health* **49,** 903.
Orlob, G. B. (1962). *Phytopathology* **52,** 747.
Orlob, G. B. (1963). *Virology* **21,** 291.
Oster, G., and McLaren, A. D. (1950). *J. Gen. Physiol. Ser.* **33,** 215.
Perdrau, J. R., and Todd, C. (1933a). *Proc. Roy. Soc., Ser. B* **112,** 288.
Perdrau, J. R., and Todd, C. (1933b). *Proc. Roy Soc., Ser. B* **112,** 277.
Santamaria, L. (1960). *Boll. Chim. Farm.* **99,** 464.
Sastry, K. S., and Gordon, M. P. (1966). *Biochim. Biophys. Acta* **129,** 49.
Schaffer, F. L. (1962). *Virology* **18,** 412.
Schaffer, F. L., and Hackett, A. J. (1963). *Virology* **21,** 124.
Scholtissek, C., and Rott, R. (1964). *Nature (London)* **204,** 39.
Sharp, D. G., and Kim, K. S. (1966). *Virology* **29,** 359.
Shaughnessy, H. J., Wolf, A. M., Janota, M., Neal, J., Oppenheimer, F., Milzer, A., Naftulin, H., and Morrissey, R. A. (1957). *Proc. Soc. Exp. Biol. Med.* **95,** 251.
Shortt, H. E., and Brooks, A. G. (1934). *Indian J. Med. Res.* **22,** 529.
Simon, M. I., and Van Vunakis, H. (1962). *J. Mol. Biol.* **4,** 488.
Singer, B., and Fraenkel-Conrat, H. (1966). *Biochem.* **5,** 2446.
Sinkovics, J. G., Bertin, B. A., and Howe, C. D. (1965). *Cancer Res.* **25,** 624.
Smith, K. C., and Hanawalt, P. C. (1969). "Molecular Photobiology: Inactivation and Recovery." Academic Press, New York.
Spikes, J. D. (1968). *In* "Photophysiology" (A. C. Giese, ed.), Vol. 3, p. 33. Academic Press, New York.
Stair, R., and Johnston, R. G. (1954). *J. Res. Nat. Bureau Standards* **53,** 211.
Thormar, H., and Peterson, I. (1964). *Acta Pathol. Microbiol. Scand.* **62,** 461.
Tomita, Y., and Prince, A. M. (1963). *Proc. Soc. Exp. Biol. Med.* **112,** 887.
Uber, F. M. (1941). *Nature (London)* **147,** 148.
Wallis, C., and Melnick, J. L. (1963). *Virology* **21,** 332.
Wallis, C., and Melnick, J. L. (1964). *Virology* **23,** 520.
Wallis, C., Sakurada, N., and Melnick, J. L. (1963). *J. Immunol.* **91,** 677.
Waskell, L. A., Sastry, K. S., and Gordon, M. P. (1966). *Biochim. Biophys. Acta* **129,** 49.
Welsh, J. N., and Adams, M. H. (1954). *J. Bacteriol.* **68,** 122.
Wilson, J. N., and Cooper, P. D. (1962). *Virology* **17,** 195.
Wilson, J. N., and Cooper, P. D. (1963). *Virology* **21,** 135.
Yamamoto, N. (1958). *J. Bacteriol.* **75,** 443.
Yamamoto, N., Hiatt, C. W., and Haller, W. (1964). *Biochim. Biophys. Acta* **91,** 257.
Zelle, M. R., and Hollaender, A. (1954). *J. Bacteriol.* **68,** 210.

Chapter 3

The Effects of Ultraviolet Radiations on Bacteria

BARBARA ANN HAMKALO

I. Ancient History

Radiations at wavelengths from 190 to 380 nm comprise the near (300–380 nm)- and far (190–300 nm)-ultraviolet (UV) regions of the electromagnetic spectrum. The bactericidal effects of such radiations were first noted nearly one hundred years ago when Downes and Blunt (1878) reported bacterial killing by sunlight. With the use of the carbon arc spectrum, Ward (1894) achieved maximal killing of anthrax spores in the spectral region from the blue into the far UV. Many reports followed these which detailed the bactericidal effects of UV (Duggar, 1936, for review). However, the site and mechanism of UV-induced inactivation eluded investigators until Henri (1914) cited the absorption of UV by the nucleus as the event which leads to cellular inactivation. He also suggested that mutations ("heritable modifications") in *Staphylococcus aureus* might result from sublethal doses of UV.

Although Henri's observations are significant, most of the work reported up to the first half of the 20th century did not advance the field of radiation biology significantly because of the lack of quantitation of results, nonuniformity of doses, use of nonmonochromatic light and a lack of knowledge of the effects of pre- and postirradiation treatments upon colony-forming ability of bacteria (Zelle and Hollaender, 1955). Gates (1928, 1929) published a series of well-controlled experiments in which he determined the action spectrum of UV killing of bacteria. Figure 1 illustrates the amount of light at different wavelengths required to kill 50% of a culture of *Escherichia coli*. In these experiments, Gates used monochromatic light isolated by quartz prisms; he measured incident energies and estimated absorption by cells.

Since the maximum efficiency of inactivation occurs at 260 nm, and since pyrimidine residues of nucleic acids absorb maximally at the same wavelength, Gates concluded that absorption of UV by nucleic acids is a lethal event. Action spectra for the UV-inactivation of a variety of microorganisms were subsequently reported (Hollaender and Claus, 1936; Loofbourow, 1948); these data focused attention and further research upon how and why cells are killed as a result of the absorption of UV by nucleic acids.

In an attempt to prove conclusively the primary target of UV-induced lethality, Kelner (1953) studied the effects of 254 nm light upon DNA and RNA syntheses, growth, and respiration in *E. coli* B/r, a radiation-resistant strain (see Section IV,B). He observed immediate inhibition of DNA synthesis after UV, while other processes showed little or delayed inhibition. However, this inhibition of DNA synthesis could be

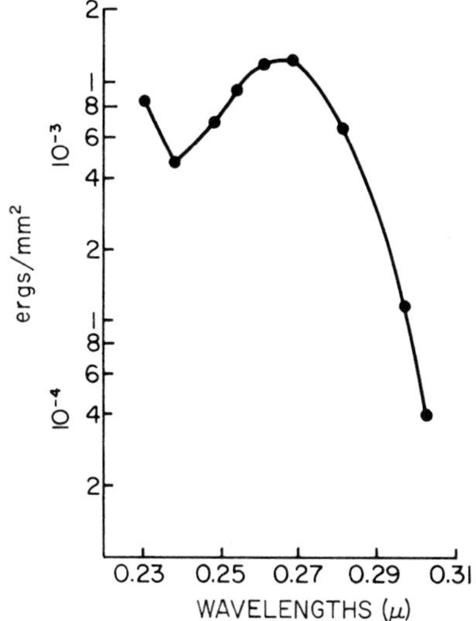

Fig. 1. Action spectrum for the inactivation of a culture of *E. coli.* (Data from Gates; adapted by Zelle and Hollaender, 1955.)

reversed by post-UV illumination with long-wavelength light (photoreactivation; see Section IV,C,4,a).

On the basis of these observations, Kelner developed a model to relate the following series of events which ensue after irradiation of bacteria. UV inhibits nuclear functioning; DNA synthesis ceases immediately after a dose of UV which does not markedly effect RNA synthesis; cytoplasmic reactions continue for a while although the inhibition of DNA synthesis ultimately affects all nuclear-governed processes.

The following sections of this chapter deal with experimental data gathered over the past fifteen years that elucidate the mechanism of UV action in bacteria and elaborate upon and substantiate the Kelner hypothesis.

II. The Site of UV-Induced Inactivation: Changes in Polynucleotides

A. Pyrimidine Dimers

In the past ten years, attention has turned to the identification of photoproducts in UV-irradiated synthetic polynucleotides or nucleic acids isolated from cells. A major achievement in these investigations was reported by a group of photochemists, Beukers *et al.* (1958), who discovered that upon UV-irradiation at 254 nm, a frozen solution of thymine reversibly loses the A^{264} characteristic of thymine. Reversal can be effected by reirradiation at shorter wavelengths (240 nm). They interpreted these results as the formation of a photoreversible, UV-induced product that contains thymines fixed by freezing in a favorable orientation for the photochemical reaction. By elemental analysis, crystallography, and infrared spectroscopy, Beukers and Berends (1960) identified the photoproduct as a dimer of two adjacent thymines which are linked by a cyclobutane ring through their 5- and 6-position carbon atoms ($\widehat{TT}$). Subsequent experiments proved that a steady state exists between dimer formation and reversal. The formation of dimers is favored by long wavelength irradiation (280 nm), and monomerization is favored at shorter wavelengths (240 nm).

The helical structure and base-stacking of DNA offer the possibility that adjacent thymines will be in favorable orientation for dimerization. Subsequent work focused upon the identification of UV-induced photoproducts in native DNA. However, in order to implicate pyrimidine dimers as those photoproducts most responsible for biological inactivation, certain criteria have to be met: The photoproducts should be isolable from the nucleic acids of irradiated cells and doses required to produce these photoproducts must be in the range of doses used to cause biological inactivation.

Thymine dimers in fact have been found in UV-irradiated DNA by Wacker *et al.* (1960). R. B. Setlow and Carrier (1963) showed that thymine dimers formed in DNA by long-wavelength irradiation (280 nm) can be reversed by subsequent short-wavelength irradiation (240 nm). Dimers can be isolated from DNA by acid hydrolysis (usually 88% formic acid) followed by paper chromatography (butanol–acetic acid–water), since dimers have an R_f distinct from thymine (Smith, 1963).

R. B. Setlow and Setlow (1962) provided definitive evidence that the frequency of pyrimidine dimers after UV is closely correlated with the degree of inactivation. Based on the wavelength dependence of dimer formation and reversal, they irradiated transforming DNA isolated from *Hemophilus influenzae* at 280 nm, assayed for transforming activity, then reirradiated the DNA at 240 nm, and again assayed for transformation. Figure 2 shows that increases in the dose of dimer-producing irradiation decrease the transforming activity and monomerization of the dimers restores transformation.

Other evidence points to the role of pyrimidine dimers in cellular inactivation: Calculations by R. B. Setlow (1964b) show that the quantities of dimers in irradiated bacteria, viruses, or extracted DNA are sufficient to account for UV-inactivation (see Table I); dimers in primer

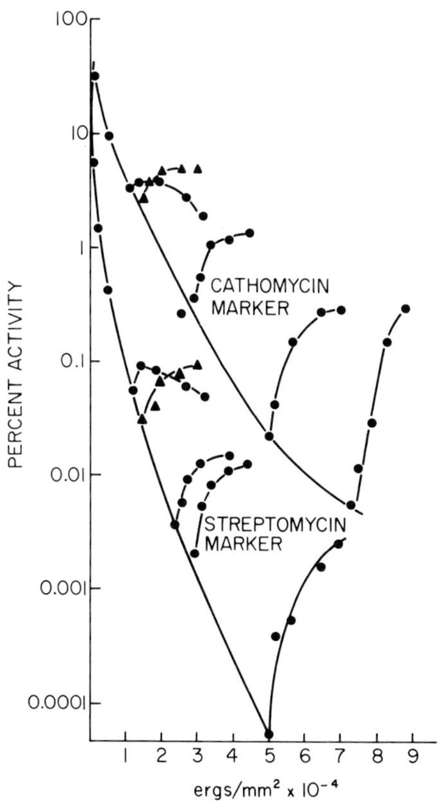

Fig. 2. Transforming activity for two markers as a function of dose at 280 nm, followed by reirradiation at 240 nm. (From R. B. Setlow and Setlow, 1962.)

DNA interfere with *in vitro* polymerization (Bollum and Setlow, 1963); dimers act as blocks to DNA synthesis *in vivo* (R. B. Setlow *et al.*, 1963; see Section VI,A); dimers are removed from the DNA of radiation-resistant organisms but not from the DNA of radiation-sensitive organisms (R. B. Setlow and Carrier, 1964; see Section IV,A); photoreactivating (PR) illumination in the presence of the photoreactivating enzyme results in the disappearance of dimers from DNA and a concomitant increase in biological activity (J. K. Setlow and Setlow, 1963; Rupert, 1964; J. K. Setlow, 1966, 1967, for reviews; see Section V,C,2).

B. Other Photoproducts

Many other pyrimidine-containing photoproducts are formed upon UV-irradiation of polynucleotides. Pyrimidine dimers of uracil $(\widehat{UU})$, cytosine ($\widehat{\mathrm{CC}}$), uracil–thymine ($\widehat{\mathrm{UT}}$), cytosine–thymine ($\widehat{\mathrm{CT}}$) and uracil–cytosine ($\widehat{\mathrm{UC}}$) are isolable *in vitro*. However, with the exception of $\widehat{\mathrm{CT}}$, little is known of the role of these photoproducts in biological inactivation (Smith, 1966). Ono *et al.* (1965) have shown that UV-irradiated polycytidylic acid exhibits heat-reversible altered transcription by RNA polymerase from the incorporation of GTP to that of ATP. These observations implicate cytosine hydrate formation in the altered transcriptive activity.

Thymine-containing photoproducts other than cyclobutane-type dimers appear to be responsible for UV-induced lethality of *Bacillus megaterium* spores irradiated at 25°C (Donnellan and Setlow, 1965). Chromatographically identical products have also been found by Rahn and Hosszu (1968) in isolated DNA and in vegetative cells irradiated at —80°C. Irradiation of spores at —80°C also results in an increase in the formation of these photoproducts (Donnellan *et al.,* 1968). Conformational differences in the DNA of spores and vegetative cells at low temperatures as compared to the DNA of vegetative cells at room temperature could explain the enhancement of spore photoproduct formation (Stafford and Donnellan, 1968). These results may be complicated by DNA–protein cross-link formation which occurs in *E. coli* B/r cells irradiated at —79°C. Cross-link formation under these conditions has been correlated with inactivation by Smith and O'Leary (1967).

In *Micrococcus radiodurans,* a "superman" among radiation-resistant organisms, action spectra for UV-inactivation have shown equally efficient killing at 260 and 280 nm (J. K. Setlow and Boling, 1965). These results suggest that damage to nucleic acids and protein contribute equally to cell death. Moseley (1969) has isolated two radiation-sensitive mutants of *M. radiodurans* that are more sensitive to 260 nm radiations than to 280 nm. In these mutants, differences in the ability to remove $\widehat{\mathrm{CT}}$ dimers appear to correlate with sensitivity.

Alterations in pyrimidine residues of RNA as a result of UV-irradiation have not been as thoroughly investigated as the corresponding changes in DNA. However, one can say that RNA is susceptible to photoinduced changes. Hydrated molecules of uracil and cytosine are heat- and acid-reversible UV-induced photoproducts which may play a role in the inactivation of RNA. Grossman (1962, 1963) studied the coding properties of UV-irradiated polyuridylic acid in a cell-free, protein-synthesizing system. After 245 nm UV, he measured a heat-reversible decrease in phenylananine incorporation (codon = UUU), accompanied by increased serine incorporation (codon = UCU). There was no noticeable change after 254 nm irradiation. Ottensmeyer and Whitmore (1967) substantiated Grossman's results and reported that uracil dimers code as UU and GU while uracil hydrates code as C. Although hydrates and dimers may be important in the inactivation of RNA species, their existence has not yet been demonstrated in RNA isolated from cells, although a specific bacterial messenger RNA (β-galactosidase) has been shown to lose activity after UV (Swenson and Setlow, 1964; see Section V,B).

III. The Fate of Pyrimidine Dimers in DNA

A. Excision

Pyrimidine dimers in DNA appear to be responsible for a major fraction of UV-induced inactivation (Table I), and the inhibition of DNA synthesis seen immediately after irradiation of *E. coli* B/r (Kelner, 1953; R. B. Setlow *et al.*, 1963) is a consequence of this damage. The macromolecular basis of UV sensitivities of different strains of *E. coli* can be understood by analyses of the fates of pyrimidine dimers in the DNA's of radiation-resistant and sensitive organisms. One model for the repair of damaged DNA that is frequently quoted is summarized by R. B. Setlow (1967). The steps in the "cut-and-patch" model include: (1) enzymatic excision of pyrimidine–dimer-containing regions of the DNA by endo- and exonuclease activities, (2) replacement of gaps by polymerization utilizing a DNA polymerase, and (3) rejoining of newly synthesized regions to the polynucleotide backbone by ligase activity.

If *E. coli* B/r is labeled with ^{3}H-thymidine, UV-irradiated (254 nm), and incubated in the dark, there is a loss of radioactivity in the acid-insoluble fraction of cell extracts (polynucleotides) with a concomitant increase in the acid-soluble fraction (oligonucleotides) (R. B. Setlow and Carrier, 1964). These results were interpreted as the excision of thymine–

TABLE I

Relative Efficiency of Production of Damage in DNA[a]

Damage	Dose (ergs/mm^2) at 254 nm to produce an average of one lesion in *E. coli* DNA
Chain breaks	~400
DNA cross-links	~400
DNA–protein cross-links	~60
Cytosine hydrates in denatured DNA	~0.2
Thymine dimers	~0.1

[a] From J. K. Setlow (1967).

dimer-containing regions in this radiation-resistant strain. Although an average of thirty nucleotides were removed for each dimer excised, dimers were found in tri- or tetranucleotides; the size of oligonucleotides may be reduced by nuclease activity during extraction. Similar experiments performed with a sensitive strain, B_{s-1}, showed no loss of radioactivity from the acid-insoluble fraction after the same incubation period. The interpretation of the results is straight-forward: B/r possesses a dark-repair system which is responsible for the removal of pyrimidine dimers from irradiated DNA; B_{s-1} does not possess an equivalent repair mechanism. The extreme sensitivity of B_{s-1} (see Section IV,A) undoubtedly results from the lack of an excision mechanism, and subsequent breakdown of DNA, as measured by Suzuki *et al.* (1966).

The removal of pyrimidine dimers from DNA appears to be a generalized repair mechanism in radiation-resistant microorganisms. Boyce and Howard-Flanders (1964) found that dimers disappeared from the DNA of UV-irradiated *E. coli* K-12 AB1157 (radiation-resistant) but not from the DNA of AB1886 (radiation-sensitive) after both strains were incubated in the dark. However, the lack of excision does not totally account for extreme UV-sensitivity. Rupp and Howard-Flanders (1968) have shown that at least one UV-sensitive K-12 strain can cope with the presence of dimers in DNA sufficiently to exhibit UV resistance which is greater than B_{s-1} (see Section III,B).

R. B. Setlow (1964a) has summarized the known conditions and requirements for excision. Although net DNA, RNA, and protein syntheses are not required, a small amount of repair replication presumably occurs to fill in gaps (see Section III,B). The process is energy-requiring since dimers are removed very slowly in the absence of a carbon source. Chloramphenicol after UV does not inhibit excision in B/r; therefore the enzyme(s) required are normally present in these cells. Incubation of irradiated cells with acriflavine of caffeine severely inhibits excision and increases inactivation by an undetermined mechanism.

Carrier and Setlow (1968), Takagi *et al.* (1968), and J. C. Kaplan *et al.* (1969) have independently isolated endonucleases from extracts of *Micrococcus luteus* that are specific for UV-irradiated native DNA and that produce single-stranded nicks in the DNA in numbers approximately equivalent to the numbers of pyrimidine dimers produced by the dose utilized. Kelly *et al.* (1969) studied the 5′ to 3′ exonuclease activity of *E. coli* DNA polymerase I on mismatched regions of DNA, which arise from local distortions in the polynucleotide chains. They propose that a pyrimidine dimer is a mismatched region which results in reduced hydrogen-bonding properties of a few base pairs. After endonucleolytic nicking, a small region of the DNA surrounding a dimer might then be

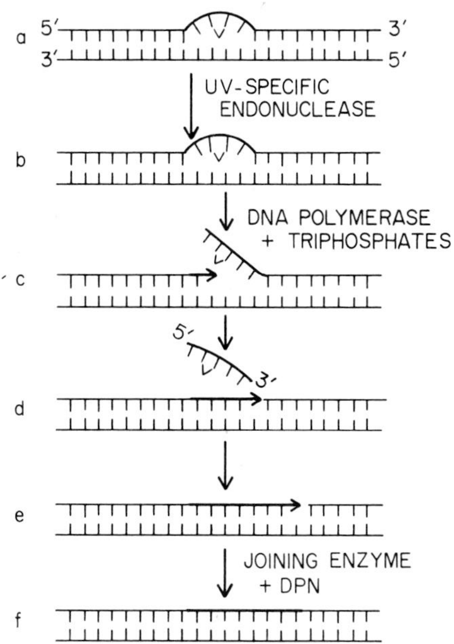

Fig. 3. A model for the repair of pyrimidine–dimer-containing DNA. (From Kelly *et al.,* 1969.)

susceptible to the 5′–3′ exonuclease. A substrate of nicked, UV-irradiated *E. coli* DNA or $p(T)_{200}$ is indeed more susceptible to this exonuclease when compared to unirradiated DNA. Figure 3 schematically represents the potential role of polymerase I in cut-and-patch repair. Although *pol I*⁻ mutants are more sensitive to UV irradiation and the rate of repair of single-strand breaks after excision is slower than in the parental strain, the mutants appear to excise dimers normally (Boyle *et al.,* 1970).

J. C. Kaplan *et al.* (1969) isolated an exonuclease activity from *M. luteus* extracts that, in conjunction with the endonuclease specific for UV-irradiated DNA, removes dimer-containing regions. The specific UV endo- and exonuclease activities may accomplish excision with polymerase I merely by filling in gaps.

The process of dimer excision is significant in the removal of damaged regions of DNA in many bacterial systems. It is responsible for host cell reactivation (*hcr*); that is, repair of irradiated bacteriophage DNA by host enzymes. (R. B. Setlow, 1966). Certain phage possess genetic information for enzymes required to repair their own irradiated DNA by excision after infection of *hcr*⁻ hosts (*v*-gene repair of T4) (Harm, 1959; R. B. Setlow and Carrier, 1966). Vegetative *Bacillus subtilis* and *B. megaterium* possess an excision system (Shuster, 1967). Wild-type

M. radiodurans, discussed briefly in Section II,A, appears to be able to excise both $\widehat{TT}$ and $\widehat{CT}$ dimers extremely rapidly (Boling and Setlow, 1966), while UV-sensitive mutants of *M. radiodurans* appear to excise $\widehat{CT}$ dimers at a decreased rate (Moseley, 1969). Since the DNA of this microorganism has a high GC content, cytosine-containing dimers become quite significant in inactivation.

B. Replacement of Excised Regions

Pettijohn and Hanawalt (1964) demonstrated a small amount of "repair synthesis" in *E. coli* TAU-bar after high UV doses. A density label (5-bromouracil) was dispersively and randomly incorporated into small regions of both strands of the DNA during the period of inhibition of net DNA synthesis. There is, however, no definitive evidence of the location of incorporated label. Also, more label was incorporated than one would expect if only regions of the bacterial chromosome containing dimers were removed.

Weiss and Richardson (1967), Hurwitz *et al.* (1967), and Gellert (1967) characterized a class of enzymes isolated from bacteria and phage-infected cells that possess the ability to catalyze phosphodiester bond formation. Such enzymes may be utilized in rejoining the repaired regions of DNA to the preexisting polynucleotide chains.

The fact that recombination-deficient (*rec*⁻) mutants of *E. coli* are UV-sensitive (Clark and Margulies, 1965) suggests that recombination may share some components of the excision system.

IV. Measurements of Radiation Sensitivity

A. Survival Curves

Survival is a measurement frequently utilized to assess the radiation sensitivity of microorganisms. The information one can derive from such data, however, is limited. As will be discussed in this section, survival measurements can be altered by growth conditions, plating conditions, and pre- and posttreatment of cells.

The surviving fraction of cells and the shape of a survival curve are used to determine the relative sensitivity of bacterial strains and, by inference, to determine to what degree cells can cope with UV-induced damage in their DNA's. The mean lethal dose (MLD, $1/e$ dose or D_{37})

is that dose required to inactivate cells to 37% survival and is a convenient measure of relative radiation sensitivities.

A comparison of the $1/e$ doses for the inactivation of several variants of *E. coli* B illustrates the utility of such a measurement. *Escherichia coli* B is inactivated to 37% survival after 500 ergs/mm^2; 70 ergs/mm^2 is required for the same inactivation of B/r; and 1 erg/mm^2 is sufficient to kill most B_{s-1} cells. All these doses are at 254 nm and have been taken from Swenson and Setlow (1966). Figure 4 illustrates survival curves of these three strains. Hill (1958) obtained these curves by 254 nm irradiation of stationary phase cultures in a nonabsorbing medium, followed by dilution and plating on nutrient agar. The obvious differences are the shapes of the curves and the relative inactivations after any given dose. B_{s-1} is exquisitely sensitive; even very low doses inactivate a culture to less than 1% survival. Strain B appears intermediate in sensitivity and B/r is clearly most resistant. These curves substantiate the sensitivities based on the $1/e$ doses, although the latter figures represent survivals of log phase cultures.

The exponential decreases in survival of *E. coli* B and B_{s-1} are consistent with one-hit kinetics—a single event is lethal; the sigmoidal sur-

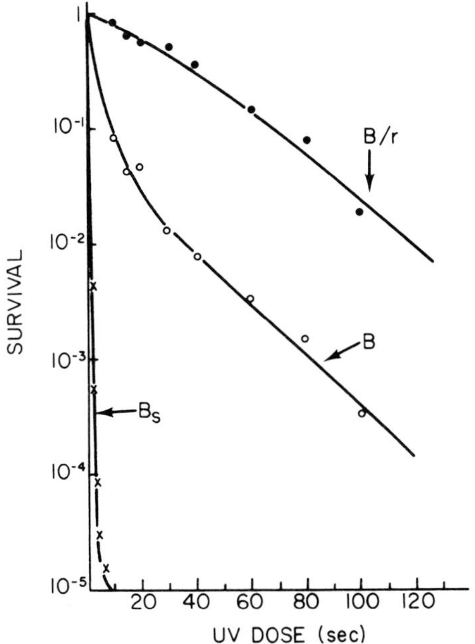

Fig. 4. Ultraviolet survival curves of stationary phase *E. coli* strains B, B/r, and B_{s-1}. (From Hill, 1958.)

vival curve characteristic of B/r and other radiation-resistant strains indicates multihit kinetics—that is, more than one event is required for inactivation. After 254 nm irradiation at doses in the region of the shoulder of the B/r survival curve (Takebe and Jagger, 1969) growth and division delays are measurable. These delays may increase the efficiency of repair, resulting in less lethality than expected at these low doses.

Since R. B. Setlow and Carrier (1964) noted that B/r loses dimers from its DNA in the dark while B_{s-1} cannot remove dimers, it appears that the differences in sensitivity are at least partly due to differences in excision proficiency. An examination of the genetic differences among the three variants described above confirms this interpretation and permits a possible explanation for the sensitivity of *E. coli* B as compared to B/r.

B. Genetics of Sensitivity

Witkin (1946, 1966a,b), Hill (1958, 1965), and Howard-Flanders *et al.* (1964, 1966) have been instrumental in the isolation and analysis of various mutants of *E. coli* B and K-12 that differ in UV sensitivities. Witkin (1946) isolated B/r from *E. coli* B by selection of viable cells after 1000 ergs/mm^2 of 254 nm UV. Hill (1958) and Hill and Simson (1961) isolated and characterized B_{s-1} and several other UV-sensitive mutants from heavily irradiated cultures of strain B.

Although both *E. coli* B and B/r possess the ability to remove pyrimidine dimers from their DNA's, B is more sensitive than B/r. After extremely low UV doses (35 ergs/mm^2), strain B forms long, filamentous cells. Van de Putte *et al.* (1963) mapped the locus that controls filament formation (*fil*$^-$), and Deering and Setlow (1957) found a nucleic acid action spectrum for filament formation. This physiological defect is also characteristic of certain K-12 strains (*lon*$^-$) and appears to be caused by defective cell plate formation in the presence of UV-induced damage to DNA (Adler and Hardigree, 1964; Howard-Flanders *et al.*, 1964). It is this mutation that appears to be responsible for the greater UV sensitivity of B compared to B/r at low doses.

Escherichia coli B_{s-1} has been shown by Greenberg (1967) and Mattern *et al.* (1966) to differ from B/r at two loci, designated *hcr*$^-$ (inability to host cell reactivate) and *exr*$^-$ (inability to excise pyrimidine dimers). These mutations additively determine the sensitivity of B_{s-1}, although the precise difference between *hcr*$^-$ and *exr*$^-$ is not entirely clear.

The genetic differences which determine radiation sensitivity in *E. coli*

B are paralleled by mutants of *E. coli* K-12. Howard-Flanders *et al.* (1966) isolated radiation-sensitive mutants which are able to plate phage T1 at a normal efficiency but which plate UV-irradiated T1 at a very low efficiency (hcr^-). These mutants are unable to excise pyrimidine dimers from their DNA's and map at three distinct loci, *uvrA, uvrB,* and *uvrC* (Howard-Flanders, 1968). All loci are phenotypically identical, and there is no evidence for sequential action of gene products.

Although there are differences in the way *E. coli* B and K-12 mutants behave after UV (see Section V,A), the genetic differences in both strains are expressions of the capacity of cells to repair or cope with the presence of UV-induced damage in their DNA's.

C. Treatments Affecting Inactivation

1. *Growth Phase of Cells*

The survival of a culture of irradiated cells appears to depend upon many criteria in addition to the genetically controlled proficiency of cells to repair UV-induced dimers, since alterations in cultural conditions before and after irradiation can alter the apparent efficiency with which a cell copes with UV-induced lesions. Stationary phase cells are more resistant to irradiation than are log phase cells (Woodside *et al.*, 1960). Ginsberg and Jagger (1965) concluded that the amount of UV-induced lethal damage is independent of the growth phase, based upon the measurement of similar photoreactivable sectors (see Section IV,C,4,b) of killing in both log and stationary phase cultures of *E. coli* 15 $T^-A^-U^-$. Changes in sensitivity with growth rate have been correlated with changes in the number of nuclei (genomes) per cell (Yanagita *et al.*, 1958), although Witkin (1951) did not find a positive correlation of resistance with nuclear number.

Hanawalt (1966) studied changes in resistance (as measured by survival) in replicating and nonreplicating cultures of *E. coli* TAU (exr^+) and B_{s-1} (exr^-). Reinitiation of replication was prevented by withholding arginine and uracil from TAU and chloramphenicol treatment of B_{s-1}. No differences were observed in the survival curves of B_{s-1} irradiated during a round of replication or after completion of a round of replication, whereas nonreplicating exr^+ cells are more radiation resistant. Hanawalt interprets the increased resistance of nonreplicating (increased shoulder of the survival curve) as an indication of increased time available for repair synthesis before semiconservative replication begins to compete with the excision process.

Helmstetter and Uretz (1963) found an increase in the survival of

E. coli B about halfway through the division cycle of synchronously growing cultures. Since Cooper and Helmstetter (1968) showed that cells growing under similar conditions exhibit gaps between rounds of replication, a gap midway in the division cycle could also provide increased time for repair.

Since net DNA synthesis is inhibited immediately after UV in most resistant strains (Kelner, 1953; R. B. Setlow *et al.*, 1963), it is not entirely clear why a resting chromosome is more efficiently or effectively repaired compared to a replicating chromosome. It is possible that a growing point is more susceptible to attack by nucleases during repair, and therefore a chromosome that lacks a growing point might be less likely to sustain additional damage during the repair reactions. Another pertinent question, which will be discussed in Section V,A, is the relationship among the amount of damage that remains at the time semiconservative replication resumes, the rate of replication upon resumption, and the biological activity of the DNA made in the post-repair period.

The extreme case of changes of UV resistance to UV during the cell cycle is seen in bacterial spores which are six to ten times as resistant as vegetative cells (Romig and Wyss, 1957).

2. *Plating Media*

The medium upon which irradiated cells are plated may result in altered survival curves. One example of this kind of alteration is seen in Fig. 5, which shows survival of log phase *E. coli* B/r, grown on either glucose or glycerol, irradiated, and plated on either nutrient agar (a) or synthetic media (b) plates. Plating on synthetic media decreases the survival an order of magnitude after any dose and abolishes the shoulder of the survival curve. These data are in agreement with Alper and Gillies (1960) and might be interpreted as a low efficiency of repair in cells plated on synthetic medium. The question of the relative efficiencies of repair in suspension and on plates, however, has not yet been studied in detail.

3. *Pretreatments*

a. Growth Prior to UV. Stapleton and Engel (1960) found increases in the shoulders of survival curves of UV-irradiated *E. coli* B/r cells grown on glucose–peptone as compared to cells grown on peptone prior to UV. The increased resistance of glucose-grown cells, noted in Section V, probably reflects metabolic differences in these cells after irradiation that may facilitate more effective and efficient repair of lethal lesions.

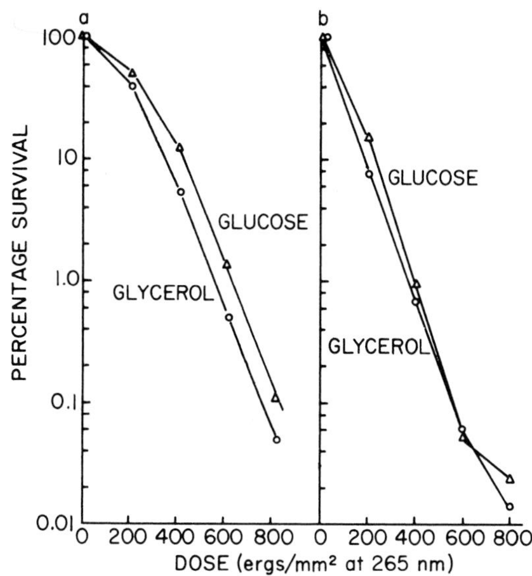

Fig. 5. Survival of *E. coli* B/r after 265 nm irradiation, followed by plating on nutrient agar (a) or synthetic medium (b) plates. (From Hamkalo and Swenson, 1969.)

b. Near-UV Illumination. If log-phase cultures of *E. coli* are illuminated with sublethal doses of near-UV light, (greater than 300 nm), delays in growth and cell division are induced. Hollaender (1943) first reported such delays and the phenomenon has since been defined as photoprotection (PP), because Jagger *et al.* (1964) found that cells irradiated with near-UV, followed by far-UV, exhibit less inactivation for a given far-UV dose. From the action spectra of these delays, which peak at 340 nm, Jagger *et al.* (1964) suggested that vitamin K_2 is the chromophore. Near-UV destruction of this vitamin, part of the electron transport chain of *E. coli,* can account for inhibition of respiration and subsequent growth and division delays (Jagger and Takebe, 1968).

The growth and division delays after far-UV irradiation could also provide additional time for the repair of damaged DNA before net syntheses resume. In addition to survival, Kantor and Deering (1967) have demonstrated photoprotection of filament formation in *E. coli* B by 335 nm UV prior to 254 nm irradiation.

4. *Posttreatments*

a. Liquid Holding Recovery. Liquid holding recovery (LHR) is a phenomenon first described by Hollaender and Claus (1937) and named and studied further by Roberts and Aldous (1949). LHR is defined as increased survival of organisms, after a given UV dose, upon holding for a variable period of time in liquid prior to plating. Figure 6 shows that survival of *E. coli* B can be increased 100-fold as a result of holding. Kantor and Deering (1967) demonstrated that LHR reduces filament formation in strain B; and Swenson and Setlow (1966) showed LHR of post-UV DNA synthesis in B/r, but no effect of holding on DNA synthesis after UV of B_{s-1}. It has been suggested by Harm (1966) that LHR occurs by a mechanism similar to that of PP: induction of growth and division delays that provide time for repair of DNA.

An observation by Roberts and Aldous (1949) that the composition of the holding medium is relatively unimportant for the occurrence of LHR suggests another mechanism for this effect. They found increased survival in such diverse media as distilled water and nutrient broth. It is therefore possible that the simple operation of keeping cells in suspension for a time before plating increases repair on plates (see Section IV,C,2).

b. Photoreactivation. Survival can be increased after UV if cells are illuminated with long-wavelength (330–500 nm) light before plating. This

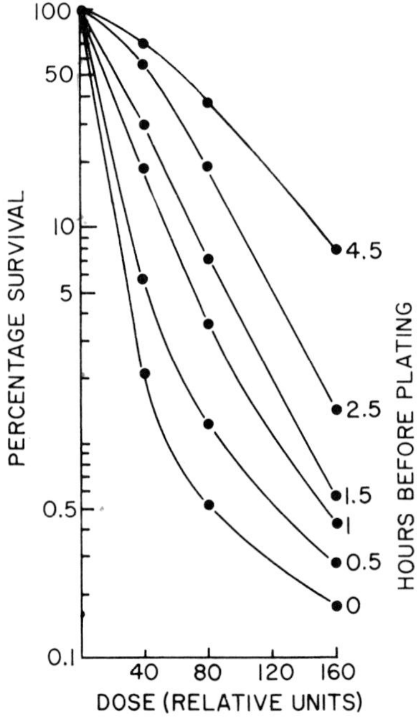

Fig. 6. Increase in survival of *E. coli* B as a function of holding times in liquid. (Data from Roberts and Aldous, 1949; redrawn by Zelle and Hollaender, 1955.)

type of reactivation has been named, appropriately, photoreactivation (PR). Kelner (1949) noted PR after illumination of UV-inactivated *Streptomyces griseus* conidia, and he subsequently made the same observations after far-UV and illumination of *E. coli* B/r (1953). Both Kelner (1949) and Novick and Szilard (1949) derived the dose reduction factor (DRF) as the effective UV dose after a maximum PR/UV dose required to give the same inactivation. Figure 7 illustrates the effect of maximum PR on the survival of *E. coli*. The DRF is independent of UV dose, and the amount of light required for maximum PR is a linear function of the inactivating dose.

Numerous UV-induced events are photoreactivable. These include the delay in DNA synthesis (R. B. Setlow *et al.*, 1963), filament formation (Kantor and Deering, 1967), delay in cell division (Kelner, 1953), mutation (Witkin, 1963), induction of prophage (Latarjet, 1951), and delay in induced enzyme formation (S. Kaplan *et al.*, 1953; Swenson, personal communication).

The temperature coefficient of PR is between 1.5 and 3.0, which suggests an enzymatic reaction. Rupert *et al.* (1958) found that extracts from photoreactivable organisms (*E. coli* and yeast) are capable of restoring activity to UV-irradiated *H. influenzae* transforming DNA if

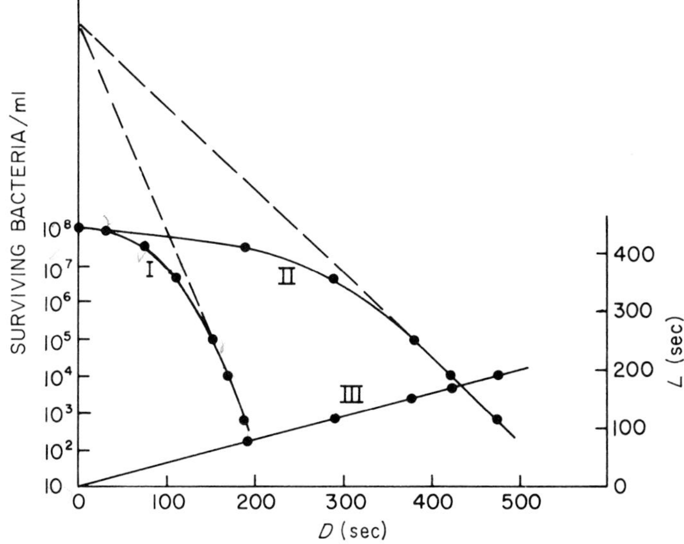

Fig. 7. Effect of photoreactivating illumination (PR) upon UV-induced inactivation in *E. coli* B/r. Curve I is dark recovery; II after maximum PR; III is UV dose after PR as a function of dose after UV alone. (Data from Novick and Szilard, 1949; from Dulbecco, 1955.)

extract plus DNA are illuminated with PR light. Muhammed (1966) achieved a 3000-fold purification of the yeast PR enzyme, but attempts to identify the chromophore have been unsuccessful. However, the possibility of an enzyme–DNA complex that absorbs the PR light has not been ruled out.

Since dimers disappear from DNA after PR treatment, J. K. Setlow and Setlow (1963) concluded that pyrimidine dimers are the photoreactivable lesions. The mechanism of PR is the direct photoenzymatic monomerization of dimers, since dimers are not found in the DNA, in the TCA-soluble fraction of cells or in the medium after PR (J. K. Setlow and Setlow, 1963). Cook (1967) has provided direct evidence that PR does occur by pyrimidine dimer monomerization. He quantitatively accounted for the disappearance of dimers by the appearance of free thymine after hydrolysis and chromatography of UV-irradiated, photoreactivated *E. coli* DNA. Jagger and Stafford (1965) defined this type of PR, or photoenzymatic PR, as "direct." They also describe "indirect PR" which is measurable in cells that do not possess a photoreactivating enzyme (phr^-). After UV irradiation at 254 nm, followed by 340 nm illumination, phr^- strains exhibit increased survival. The mechanism of indirect PR may be by induction of growth and division delays.

Although photoreactivation can reverse up to 90% of the far-UV induced lethality to bacterial cells, all inactivation is not photoreactivable. Dark repair competes for photoreactivable lesions; therefore most photoreactivable lesions are probably pyrimidine dimers (R. B. Setlow, 1966). Nonphotoreactivable lethal lesions have not been identified, although they may be important in certain types of UV-induced mutations (see Section VII).

V. Effects of Pyrimidine Dimers in DNA on Macromolecular Synthesis

The incorporation of radioactive precursors into cold trichloroacetic acid-insoluble material is routinely used as a measure of macromolecular synthesis. The filter paper disc method of Bollum (1959) is a rapid, simple, and reproducible method for such experiments. A small aliquot ($\sim$10 μl) of cells that have been incubated with the precursor is spotted onto a filter paper disc. The disc is immediately placed into cold 5% TCA for about 20 min. If 10 ml of TCA is allotted for each disc, any number of samples can be processed simultaneously. After the first TCA wash, the discs are rinsed for 10 min with fresh cold 5% TCA and dried before counting in a liquid scintillation spectrometer.

A. DNA Synthesis

As Kelner (1953) first reported, net DNA synthesis is inhibited immediately after UV-irradiation of *E. coli* B/r. With time, net DNA synthesis does resume in the resistant strain. After 254 nm irradiation of B_{s-1}, R. B. Setlow *et al.* (1963) showed that there was little, if any, net DNA synthesis after UV and no recovery of net synthesis. Figure 8 illustrates dose–response curves of the incorporation of ^{3}H thymidine into DNA (acid-insoluble material) by glucose-grown log phase cultures of *E. coli* B_{s-1} and B/r. After 25 ergs/mm^2, DNA synthesis is completely and irreversibly inhibited in B_{s-1} unless photoreactivating illumination is given immediately after UV. A similar dose (54 ergs/mm^2) given to B/r results in a short lag in thymidine incorporation, after which time net synthesis resumes at a rate parallel to unirradiated cells. The lag is photoreactivable and its duration is dose-dependent up to about 200 ergs/mm^2. At higher doses, net synthesis resumes at the same time after UV, but the rate of synthesis becomes progressively lower (Doudney, 1965). It is of interest to note that above 200 ergs/mm^2 the survival of B/r decreases exponentially.

Based on the assumption made by Swenson and Setlow in 1966 that a dimer is a block to DNA synthesis in these strains, then B_{s-1}, which

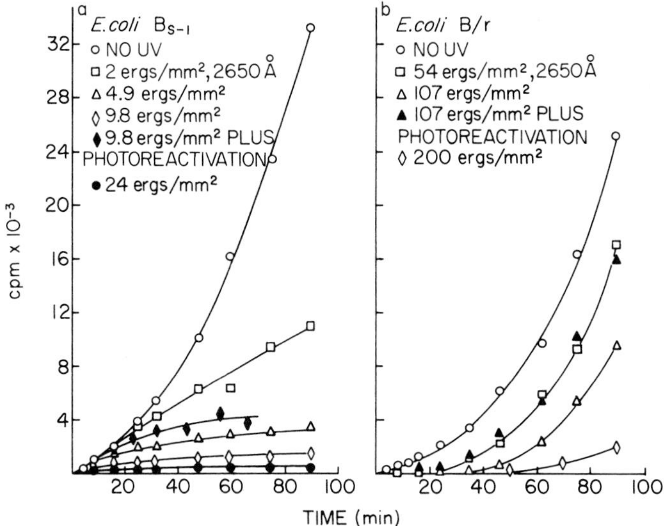

Fig. 8. Incorporation of ^{3}H-thymidine by *E. coli* B_{s-1} (a) and B/r (b) after different doses of inactivating (265 nm) and photoreactivating light. (From R. B. Setlow *et al.*, 1963.)

cannot excise dimers, can only synthesize DNA to a block. With increased UV doses, a greater number of blocks are randomly introduced into the bacterial genome, and progressively less synthesis will occur before a dimer is reached.

In B/r, the dose-dependent delay is interpreted as a period of dark repair. *Escherichia coli* B also exhibits a dose-dependent delay after UV, but the duration of the lag is shorter and the rate of synthesis upon resumption is greater than in B/r (Swenson and Setlow, 1966). These results can be correlated with the fact that B excises dimers faster (R. B. Setlow and Carrier, 1964). Since B appears to be more efficient in removing dimers, why then is it more sensitive to UV (see Section IV,A)? It has been suggested that the faster rate of excision results in a greater number of nicks at any one time, which may not be ligated (Swenson and Setlow, 1966).

At doses up to 200 ergs/mm^2, most, if not all, dimers are removed before net synthesis resumes. Above 200 ergs/mm^2, interpretations of DNA synthetic patterns become complex because net synthesis resumes at the same time, while the rates of synthesis decrease with dose. These data have suggested that, under certain conditions, synthesis past a dimer occurs. That is, at high doses, although repair is not complete, resumption of net DNA synthesis is triggered by an unknown mechanism. If the probability of replication past a dimer is low, then a lower rate of net DNA synthesis might be expected (Swenson and Setlow, 1966). Hewitt and Billen (1965) found that after UV of *E. coli* 15T$^-$, the preexisting growing point is not replicated upon resumption of synthesis; but a new site appears. After 600 ergs/mm^2 only about 50% of the DNA is ever replicated. Billen (1969) reported a preference for the site of post-UV initiation near or at the fixed origin of the chromosome.

Some UV-sensitive strains that cannot excise dimers are more resistant to UV-inactivation than B_{s-1}. In *E. coli* K-12 uvrA (*exr*$^-$), Rupp and Howard-Flanders (1968) have demonstrated replication past dimers. That is, irradiated cultures incorporate label into semiconservatively synthesized DNA. Early after irradiation, newly synthesized single strands are smaller than parental material, but with continued incubation the DNA becomes as large as unirradiated material. These workers suggest that this synthesis occurs past a dimer, leaving a gap in the strand opposite the dimer region. With time, these gaps are filled in and sealed to reconstruct intact molecules. Howard-Flanders *et al.* (1969) suggested that in a *uvr*$^-$*rec*$^+$ cell, such synthesis, followed by recombination among daughter molecules, could reconstruct duplex molecules with no dimers and thus account for the greater survival of these strains when compared to double mutants, *uvr*$^-$*rec*$^-$. Witkin (1966a) isolated an *hcr*$^-$

variant of B/r (WP2) that can also apparently tolerate dimers in its DNA as judged by the multihit character of its survival curve.

Several intriguing questions remain with regard to the experiments described. An observation made by Hamkalo and Swenson (1969) indicates that, after 500 ergs/mm^2, DNA synthesis resumes at the same time in B/r cells grown on a variety of carbon sources, although the rates of synthesis upon resumption differ quite drastically. Figure 9 shows two incorporation curves for comparison. The glycerol-grown cells (Fig. 9b) resume synthesis at a much lower rate than the glucose-grown cells (Fig. 9a). The synthetic rate in (b) is characteristic of cells grown on acetate, pyruvate, or succinate, as well as glycerol. Since survival levels are identical in all cases (Hamkalo, 1968), metabolic differences after UV may be responsible for the different rates of synthesis (see Section VI).

Control of the rate of DNA synthesis upon resumption in radiation-resistant cells is another poorly understood area. The fact that, after doses of less than 200 ergs/mm^2, B/r resumes DNA synthesis at a rate parallel to unirradiated, exponentially growing cells suggests that the irradiated cells are making more DNA per unit time than are control cells. This could be accounted for by either a more rapid rate of synthesis or reinitiation of chromosomes before a round of replication has been completed (Swenson and Setlow, 1966). This phenomenon occurs in the

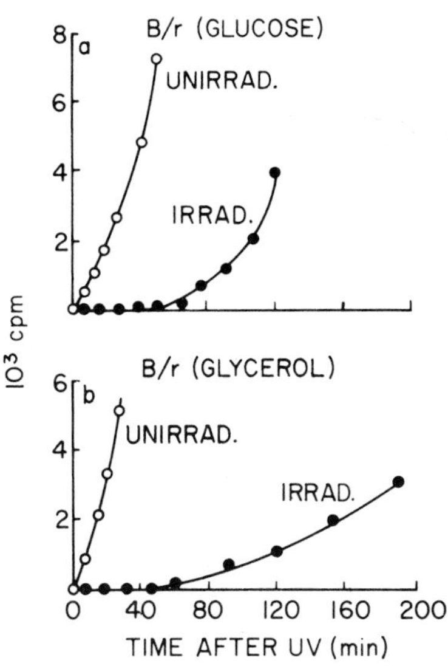

Fig. 9. Incorporation of ^{3}H-thymidine into DNA by UV-irradiated (500 ergs/mm^2) glucose-grown (a) and glycerol-grown (b) *E. coli* B/r. (From Hamkalo and Swenson, 1969.)

region of the shoulder of the survival curve and might reflect accumulation of initiation proteins in the lag period when RNA and protein are being synthesized (see Section V,B). Another alternative is an accelerated rate of DNA synthesis to catch up with other macromolecular syntheses in order to maintain a balance among macromolecular levels that might be critical for normal cell division, as suggested by Donachie *et al.* (1968) and Hamkalo and Swenson (1969). To date these suggestions are little more than pure speculation.

B. RNA and Protein Syntheses

The presence of UV-induced photoproducts in DNA undoubtedly interferes with transcription by RNA polymerase and the presence of UV-induced alterations in messenger RNA probably interferes with the messenger's ability to direct the synthesis of proteins. Doudney (1968) has provided evidence that pyrimidine dimers in DNA affect RNA synthesis. He showed that if acriflavine (an inhibitor of excision) is given to UV-irradiated B/r cells, there is a photoreactivable reduction in RNA synthesis. If the dye was administered 40 min after UV, little repression of synthesis could be measured.

Figure 10 shows net syntheses of RNA and protein after 500 ergs/mm^2 irradiation of *E. coli* B_{s-1} and B/r under different growth conditions. RNA synthesis ceases about 30 min after UV in B_{s-1} grown on glycerol (Fig. 10a) while net protein synthesis ceases at a later time (Fig. 10b). Similar results were reported for B_{s-1} cells grown on glucose (Swenson and Setlow, 1966). As first noted by Swenson and Setlow (1966), glucose-grown B/r cells resume exponential syntheses of RNA and protein at the time DNA synthesis resumes (Fig. 10, c and d). Cells grown on glycerol, however, exhibit severe inhibition of RNA and protein syntheses (Fig. 10, e and f) at the time net DNA synthesis resumes (Fig. 9b). The duration of these photoreactivable inhibitions is dose-dependent and the inhibition is correlated with a generalized inhibition of respiration and growth after UV (Hamkalo, 1968; see Section VI). Similar results are obtained for macromolecular syntheses in B/r grown on acetate, pyruvate, or succinate (Hamkalo, 1968).

Rates of RNA and protein syntheses usually reflect the amount and quality of the DNA template. In B_{s-1}, net synthesis of DNA ceases and DNA breakdown occurs after UV; therefore the cessation of RNA and protein syntheses are indeed correlated with the loss of template. In B/r, the metabolism of irradiated cells appears to alter the rate of DNA synthesis upon resumption (see Section VI and Fig. 9). Inhibition of net

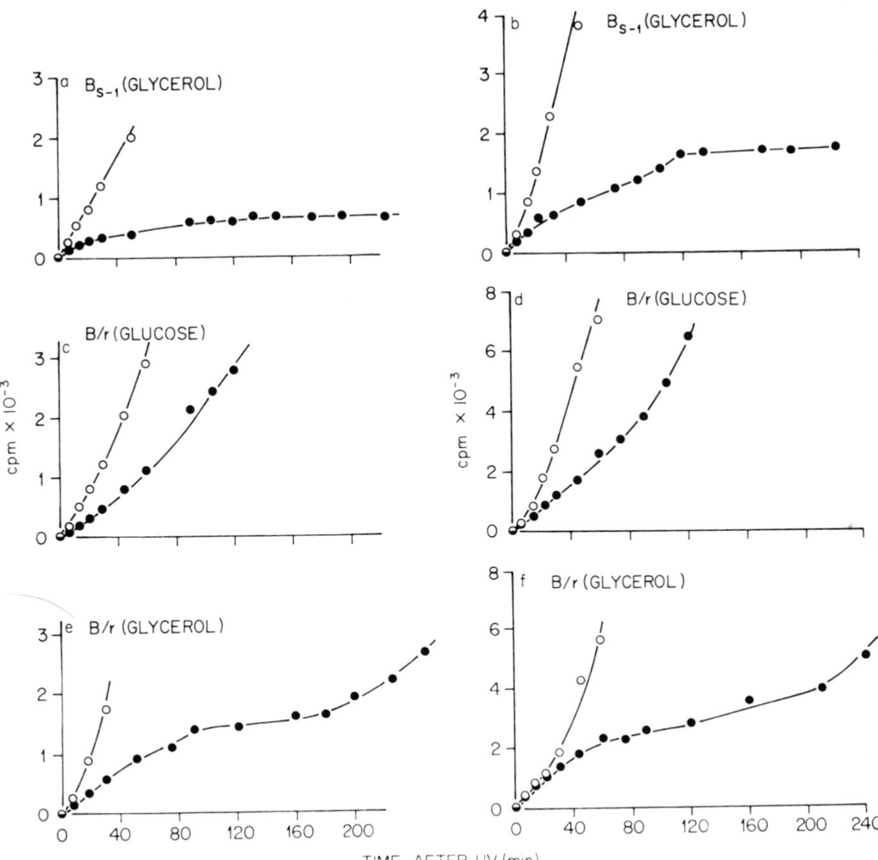

Fig. 10. Incorporation of ^{3}H-uridine and ^{3}H-leucine into UV-irradiated (500 ergs/mm^2) glycerol-grown *E. coli* B_{s-1} (a,b), glucose-grown B/r (c,d), and glycerol-grown B/r (e,f). (From Hamkalo, 1968.)

syntheses of RNA and protein occurs in cells that exhibit post-UV inhibition of respiration and growth and a low rate of DNA synthesis upon resumption. Under these cultural conditions, DNA may not be as effectively or efficiently repaired and thus provide a poor template for RNA transcription. However, almost complete inhibition of syntheses were observed in the post-repair period, and the survival of cells grown on all carbon sources studied was the same, regardless of the degree of inhibition of RNA and protein syntheses. Inhibition may reflect a cellular mechanism by which the amount of DNA per cell catches up with the quantities of other macromolecules, a requirement that may be necessary before cells can divide (Donachie *et al.*, 1968). However, Swenson and

Schenley (1970) showed that inhibition of either RNA synthesis by 5-fluorouracil (50 μg/ml) or protein synthesis by chloramphenicol (100 μg/ml) added immediately after UV prevents net inhibitions of growth and respiration, and presumably RNA and protein synthesis inhibition as well. This observation implies the UV-induced synthesis of a protein that acts to inhibit synthetic processes.

The UV-induced alterations of synthetic polyribonucleotides have been discussed in Section II. *In vitro,* UV-irradiated polydeoxyribonucleotides are transcribed by RNA polymerase differently than unirradiated polymers (Ono *et al.*, 1965), and irradiated polyribonucleotides are translated differently than unirradiated polymers in *in vitro* protein-synthesizing systems (Grossman, 1962, 1963; Ottensmeyer and Whitmore, 1967). *In vivo,* messenger RNA transcribed from a template containing dimers or messenger containing photoproducts would be expected to code for altered proteins. Induced enzyme systems have been useful in the study of UV effects on specific gene products. Although photoproducts have not yet been isolated from *in vivo* RNA induced messenger RNA for β-galactosidase can be inactivated at doses that also alter DNA (Swenson and Setlow, 1964).

Experiments directed to the question of the effects of photoproducts in DNA upon transcription are difficult to interpret because of metabolic complications. The *lac* operon is noninducible in the presence of catabolites of glucose (catabolite repression; Magasanik, 1961); therefore cells are routinely grown on glycerol in induction experiments. However, in irradiated cells, the *lac* genes are repressed by metabolite(s) of glycerol. Consequently, the presence of glycerol, an apparent sensitivity of induced messenger synthesis may in fact be caused by catabolite repression. In the absence of glycerol, Pardee and Prestidge (1967) and Swenson (personal communication) have noted a decrease in the activity of β-galactosidase immediately after UV of two different exr^+ strains of *E. coli.* The decrease is photoreactivable, and inhibitors of excision prevent recovery of inducibility. In their independent studies of recovery, Pardee and Prestidge (1967) found almost total recovery within 10 min after a high UV dose, while Swenson (personal communication) found that B/r recovery took longer than 90 min and necessitated reintroduction of glycerol. Pardee and Prestidge interpreted rapid recovery of inducibility as a selective repair of the *lac* operon. In B_{s-1}, inducibility decreased with time after UV, as did generalized RNA and protein syntheses (Swenson, personal communication).

Swenson and Setlow (personal communication) measured the near-UV inhibition of the inducibility of another enzyme, tryptophanase.

The action spectrum for inhibition peaks at 334 nm, suggesting that the chromophore is the same or similar to the absorbing component responsible for PP (see Section IV,C,3,b). The near-UV effect is sensitive to catabolite repression; in the absence of glycerol, induction is not inhibited immediately after irradiation (Swenson, personal communication).

It should be noted that in all the above experiments loss of inducibility can occur at the level of transcription and/or translation, since one merely measures the amount of protein that is sufficiently active to react in a colorimetric assay.

VI. The Physiology of Irradiated Cells

Kelner (1953) suggested that all cellular processes that depend upon functioning DNA for their activities would exhibit delayed effects of ultraviolet radiations. The delayed effects of UV extend from inhibition of RNA and protein syntheses to reductions in growth and respiration. Evidence, though somewhat contradictory, has accumulated for only slight alterations in these processes after irradiation. Reports such as those by Heinmets and Kathan (1954) that respiration in B/r is inhibited only after 10^5 ergs/mm^2 substantiate the UV resistance of oxygen consumption. However, most postirradiation physiological studies have utilized stationary phase cultures that are not active metabolically and that are highly resistant to UV as measured by survival (see Section IV,C,1). The most conclusive study on the effects of UV on metabolic processes until recently was published in 1953 by Kelner. He noted a photoreactivable decrease in the respiration rate of stationary phase cultures of *E. coli* B/r after very high doses of UV and a photoreactivable delayed inhibition of growth in log phase cultures.

Hamkalo and Swenson (1969) reported alterations in the growth and respiration of log phase cultures of B/r after relatively low doses of UV (200–500 ergs/mm^2). Cells which have been grown on glucose exhibit little post-UV inhibition of metabolism, while at the same survival level cells grown on glycerol, acetate, or pyruvate exhibit a photoreactivable, dose-dependent delayed period of severe inhibition of both growth and respiration. In all these experiments, log phase cultures (4×10^7 cells/ml) were irradiated, while being shaken in quartz Warburg vessels, by a shielded germicidal lamp (254 nm), resting at the bottom of a water bath.

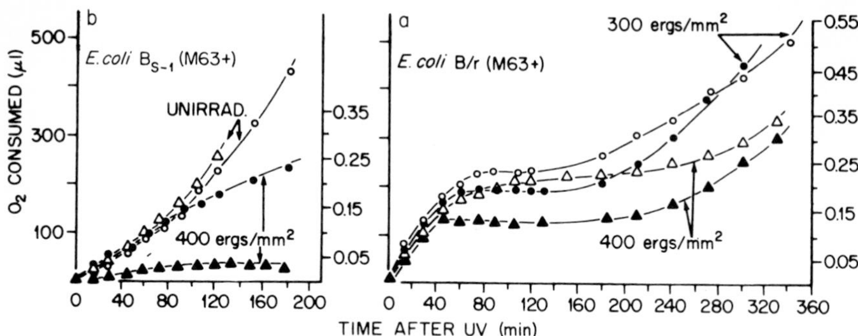

Fig. 11. Respiration (oxygen consumption) and growth (increase in A^{650}) of UV-irradiated *E. coli* B/r (a) and B_{s-1} (b). (From Hamkalo and Swenson, 1969.)

Such an experimental design permitted the measurement of respiration immediately after UV so that initial changes would be seen. In photoreactivation experiments, the germicidal lamp is replaced by a black light, and the PR illumination can be administered while the cells are incubating in the Warburg vessels. Growth is measured by the increase in turbidity at 650 nm.

Figure 11 shows post-UV growth and respiration of glycerol-grown B/r (a) and B_{s-1} (b) cells. With time after irradiation, the sensitive strain appears to lose control of coordination between normally closely coupled processes. The loss of control is interpreted as an irreversible step toward cell death as a result of the exr^- character of the strain. In fact, if one expresses the measurements in terms of oxygen consumption per unit time/increase in OD per unit time, the ratios remain fairly constant for B/r but increases to infinity in B_{s-1}.

Interpretations of the physiological changes after UV of *E. coli* B/r must be correlated with the net synthetic patterns, as discussed in Section V,B, since cellular processes do remain coupled after irradiation. The growth and respiration inhibitions are photoreactivable, but the onset of inhibition is not altered if excision is inhibited by caffeine at 1 mg/ml (Swenson and Schenley, 1970). However, if caffeine is present, respiration and growth are not resumed. Therefore, although there is no relationship between dimer excision and respiratory turnoff, dimer excision is necessary before respiration can resume. The question remains as to why severe respiratory inhibition is seen only under certain growth conditions and why resumption is related to excision. The complexities of the secondary effects of UV are beginning to be defined and the answers to these questions will undoubtedly be forthcoming.

VII. Mutation Induction—Sublethal Damage

The major portion of this review has been directed to the lethal effects of UV and the variety of changes cells may undergo in response to this damage. In addition to the damage in DNA that affects virtually all cellular processes, far-UV radiations can exert a mutagenic effect. In fact, *E. coli* B_{s-1} was isolated by Hill in 1958 as a mutant from a heavily irradiated culture of *E. coli* B. Mutation induction usually occurs after relatively low UV doses (<100 ergs/mm^2) and thus is classified as a sublethal effect.

The evidence implicating pyrimidine dimers as the lesion most responsible for inactivation of microorganisms suggests that dimers might be important as premutational lesions. Witkin (1966a) summarized lines of evidence that substantiate such a conclusion. In the presence of excision inhibitors, such as acriflavine or caffeine, the frequency of UV-induced mutations in B/r is enhanced up to 50 times. A comparison of the frequency of induced mutations in isogenic strains that differ only in excision proficiency shows virtually no mutations in the *hcr*$^+$ strain, with up to 100 times as many mutants, 90% of which are photoreactivable, in the *hcr*$^-$ strain (Hill, 1965; Witkin, 1966b).

The premutational lesion is not always a pyrimidine dimer. That is, some premutational lesions are excisable, but not photoreactivable. For example, Witkin (1966b) found no UV-induced mutations in the *lac* operon of *hcr*$^+$ cells while the same dose induced abundant, non-photoreactivable *lac* mutations in a *phr*$^+$*hcr*$^-$ strain. She suggested that the premutational lesions in this operon are excisable, but not photoreactivable, hence are not pyrimidine dimers. At other loci, up to 10% of the premutational lesions are also excisable but not photoreactivable. Also, acriflavine in the plating medium can abolish the effects of PR upon certain prototrophic mutations (Witkin, 1963).

In radiation-resistant strains (*exr*$^+$), there is a reduction in mutation to prototrophy if protein synthesis is inhibited after UV. The phenomenon is known as mutation frequency decline (MFD). Witkin (1966b) suggested that since such mutations are suppressor-type mutations (i.e., in loci that control protein-synthesizing machinery), they might result from excisable premutational lesions other than pyrimidine dimers. The inhibition of protein synthesis might increase the activity of excision enzymes toward such premutational lesions. Mutation induction by UV-irradiation appears to occur as a result of several types of lesions, including pyrimidine dimers. The occurrence of these various lesions appears to depend to some degree upon the loci under consideration. However, if lesions are not repaired, or are incorrectly repaired, these premutational lesions become the genetic bases of mutation.

VIII. Perspectives

The site and action of the absorption of ultraviolet radiations in cells have been well described by photochemical and photobiological studies over the past ten years. The problems that have been uncovered in the course of these investigations transport the study of photobiology into the realm of molecular biology. The basic principles of UV-induced photoproduct formation, reversal, and removal have been elucidated, and the genotypic and phenotypic expressions of radiation sensitivity have been explored. Future experiments will undoubtedly lead to the elucidation of the effects of the presence and removal of photoproducts in nucleic acids upon virtually every nuclear-directed cellular process, as Kelner (1953) hypothesized. The mechanisms by which photobiological changes are induced will undoubtedly be integrated into the areas of replication and recombination. The study of the effects of ultraviolet light on cells, then, appears to open the doors to understanding control, coordination, and interactions among many cellular processes.

References

Adler, H. I., and Hardigree, A. A. (1964). *J. Bacteriol.* **87,** 720.
Alper, T., and Gillies, N. (1960). *J. Gen. Microbiol.* **22,** 113.
Beukers, R., and Berends, W. (1960). *Biochim. Biophys. Acta* **41,** 550.
Beukers, R., Ijlstra, I., and Berends, W. (1958). *Rec. Trav. Chim. Pays-Bas* **77,** 729.
Billen, D. (1969). *J. Bacteriol.* **97,** 1169.
Boling, M. E., and Setlow, J. K. (1966). *Biochim. Biophys. Acta* **123,** 26.
Bollum, F. J. (1959). *J. Biol. Chem.* **234,** 2733.
Bollum, F. J., and Setlow, R. B. (1963). *Biochim. Biophys. Acta* **68,** 599.
Boyce, R. P., and Howard-Flanders, P. (1964). *Proc. Nat. Acad. Sci. U. S.* **51,** 293.
Boyle, J. M., Paterson, M. C., and Setlow, R. B. (1970). *Nature (London)* **226,** 708.
Carrier, W. L., and Setlow, R. B. (1968). *Abstr. 5th Intn. Congr. Photobiol.,* p. 19.
Clark, A. J., and Margulies, A. D. (1965). *Proc. Nat. Acad. Sci. U. S.* **53,** 451.
Cook, J. S. (1967). *Photochem. Photobiol.* **6,** 97.
Cooper, S., and Helmstetter, C. E. (1968). *J. Mol. Biol.* **31,** 519.
Deering, R., and Setlow, R. B. (1957). *Science* **126,** 397.
Donachie, W. D., Hobbs, D. G., and Masters, M. (1968). *Nature (London)* **219,** 1079.
Donnellan, J. E., Jr., and Setlow, R. B. (1965). *Science* **149,** 308.
Donnellan, J. E., Jr., Hosszu, J. L., Rahn, R. O., and Stafford, R. S. (1968). *Nature (London)* **219,** 964.

Doudney, C. O. (1965). *Symp. Fundam. Cancer Res.* **18,** 120–141.

Doudney, C. O. (1968). *Nature (London)* **219,** 1161.

Downes, A., and Blunt, T. (1878). *Proc. Roy. Soc.* **28,** 199.

Duggar, B. M. (1936). *In* "Biological Effects of Radiation" (B. M. Duggar, ed.), Vol. 2, pp. 1119–1144. McGraw-Hill, New York.

Dulbecco, R. (1955). *In Radiat. Biol.* **2,** 455–486.

Gates, F. L. (1928). *Science* **68,** 479.

Gates, F. L. (1929). *J. Gen. Physiol.* **13,** 231.

Gellert, M. (1967). *Proc. Nat. Acad. Sci. U. S.* **57,** 148.

Ginsberg, D. M., and Jagger, J. (1965). *J. Gen. Microbiol.* **40,** 171.

Greenberg, J. (1967). *Genetics* **55,** 193.

Grossman, L. (1962). *Proc. Nat. Acad. Sci. U. S.* **48,** 1609.

Grossman, L. (1963). *Proc. Nat. Acad. Sci. U. S.* **50,** 657.

Hamkalo, B. A. (1968). Ph.D. Thesis, University of Massachusetts.

Hamkalo, B. A., and Swenson, P. A. (1969). *J. Bacteriol.* **99,** 815.

Hanawalt, P. C. (1966). *Photochem. Photobiol.* **5,** 1.

Harm, W. (1959). *Z. Vererbungslehre* **90,** 428.

Harm, W. (1966). *Photochem. Photobiol.* **5,** 747.

Heinmets, F., and Kathan, R. F. (1954). *Arch. Biochem. Biophys.* **53,** 205.

Helmstetter, C. E., and Uretz, R. B. (1963). *Biophys. J.* **3,** 35.

Henri, V. (1914). *C. R. Acad. Sci.* **158,** 1032.

Hewitt, R., and Billen, D. (1965). *J. Mol. Biol.* **13,** 40.

Hill, R. F. (1958). *Biochim. Biophys. Acta* **30,** 636.

Hill, R. F. (1965). *Photochem. Photobiol.* **4,** 563.

Hill, R. F., and Simson, E. (1961). *J. Gen. Microbiol.* **24,** 1.

Hollaender, A. (1943). *J. Bacteriol.* **46,** 531.

Hollaender, A., and Claus, W. D. (1936). *J. Gen. Physiol.* **19,** 753.

Hollaender, A., and Claus, W. D. (1937). *Bull. Nat. Res. Counc.* (*U. S.*) No. 100, pp. 75–88.

Howard-Flanders, P. (1968). *Advan. Biol. Med. Phys.* **12,** 299.

Howard-Flanders, P., Simson, E., and Theriot, L. (1964). *Genetics* **49,** 237.

Howard-Flanders, P., Boyce, R. P., and Theriot, L. (1966). *Genetics* **53,** 1119.

Howard-Flanders, P., Theriot, L., and Stedeford, J. B. (1969). *J. Bacteriol.* **97,** 1134.

Hurwitz, J., Becker, A., Gefter, M. L., and Gold, M. (1967). *J. Cell. Comp. Physiol.* **70,** Suppl. 1, 181.

Jagger, J., and Stafford, R. S. (1965). *Biophys. J.* **5,** 75.

Jagger, J., and Takebe, H. (1968). *Abstr. Int. Congr. Photobiol.* 196.

Jagger, J., Wise, W. C., and Stafford, R. S. (1964). *Photochem. Photobiol.* **3,** 11.

Kantor, G. J., and Deering, R. A. (1967). *J. Bacteriol.* **94,** 1946.

Kaplan, J. C., Kushner, S., and Grossman, L. (1969). *Proc. Nat. Acad. Sci. U. S.* **63,** 144.

Kaplan, S., Rosenblum, E. D., and Bryson, V. (1953). *J. Cell. Comp. Physiol.* **41,** 153.

Kelly, R. B., Atkinson, M. R., Huberman, J. A., and Kornberg, A. (1969). *Nature (London)* **224,** 495.

Kelner, A. (1949). *Proc. Nat. Acad. Sci. U. S.* **35,** 73.

Kelner, A. (1953). *J. Bacteriol.* **65,** 252.

Latarjet, R. (1951). *C. R. Acad. Sci.* **232,** 1713.

Loofbourow, J. R. (1948) *Growth* **12,** Suppl., 77.

Magasanik, B. (1961). *Cold Spring Harbor Symp. Quant. Biol.* **26,** 249.

Mattern, I. E., Zwenk, H., and Rorsch, A. (1966). *Mutat. Res.* **3**, 374.
Moseley, B. E. B. (1969). *J. Bacteriol.* **97**, 647.
Muhammed, A. (1966). *J. Biol. Chem.* **241**, 516.
Novick, A., and Szilard, L. (1949). *Proc. Nat. Acad. Sci. U. S.* **35**, 591.
Ono, J., Wilson, R. G., and Grossman, L. (1965). *J. Mol. Biol.* **11**, 600.
Ottensmeyer, F. P., and Whitmore, G. F. (1967). *Biophys. J.* **7**, Suppl., 89.
Pardee, A. B., and Prestidge, L. S. (1967). *J. Bacteriol.* **93**, 1210.
Pettijohn, D., and Hanawalt, P. C. (1964). *J. Mol. Biol.* **9**, 395.
Rahn, R. O., and Hosszu, J. L. (1968). *Photochem. Photobiol.* **8**, 59.
Roberts, A. B., and Aldous, E. (1949). *J. Bacteriol.* **57**, 363.
Romig, W. R., and Wyss, O. (1957). *J. Bacteriol.* **74**, 386.
Rupert, C. S. (1964). *In* "Photophysiology" (A. C. Giese, ed.), Vol. 2, pp. 283–327. Academic Press, New York.
Rupert, C. S., Goodgal, S. H., and Herriot, R. M. (1958). *J. Gen. Physiol.* **41**, 451
Rupp, W. D., and Howard-Flanders, P. (1968). *J. Mol. Biol.* **31**, 291.
Setlow, J. K. (1966). *Radiat. Res., Suppl.* **6**, 144–155.
Setlow, J. K. (1967). *Compr. Biochem.* **27**, 157–208.
Setlow, J. K., and Boling, M. E. (1965). *Biochim. Biophys. Acta* **108**, 259.
Setlow, J. K., and Setlow, R. B. (1963). *Nature (London)* **197**, 560.
Setlow, R. B. (1964a). *J. Cell. Comp. Physiol.* **64**, Suppl. 1, 51–68.
Setlow, R. B. (1964b). *In* "Mammalian Cytogenetics and Related Problems in Radiobiology" (C. Pavan *et al.*, eds.), pp. 291–307. Pergamon, Oxford.
Setlow, R. B. (1966). *Science* **153**, 379.
Setlow, R. B. (1967). *Brookhaven Symp. Biol.* **20**, 1–16.
Setlow, R. B., and Carrier, W. L. (1963). *Photochem. Photobiol.* **2**, 49.
Setlow, R. B., and Carrier, W. L. (1964). *Proc. Nat. Acad. Sci. U. S.* **51**, 226.
Setlow, R. B., and Carrier, W. L. (1966). *Biophys. J.* **6**, Suppl., 68.
Setlow, R. B., and Setlow, J. K. (1962). *Proc. Nat. Acad. Sci. U. S.* **48**, 1250.
Setlow, R. B., Swenson, P. A., and Carrier, W. L. (1963). *Science* **142**, 1464.
Shuster, R. C. (1967). *J. Bacteriol.* **93**, 811.
Smith, K. C. (1963). *Photochem. Photobiol.* **2**, 503.
Smith, K. C. (1966). *Radiat. Res., Suppl.* **6**, 54–79.
Smith, K. C., and O'Leary, M. E. (1967). *Science* **155**, 1024.
Stafford, R. S., and Donnellan, J. E., Jr. (1968). *Proc. Nat. Acad. Sci. U. S.* **59**, 822.
Stapleton, G. E., and Engel, M. S. (1960). *J. Bacteriol.* **80**, 544.
Suzuki, K., Moriguchi, E., and Horii, Z. (1966). *Nature (London)* **212**, 1265.
Swenson, P. A. Personal communication.
Swenson, P. A., and Schenley, R. L. (1970). *Mutat. Res.* **9**, 443.
Swenson, P. A., and Setlow, R. B. (1964). *Science* **146**, 791.
Swenson, P. A., and Setlow, R. B. Personal communication.
Swenson, P. A., and Setlow, R. B. (1966). *J. Mol. Biol.* **15**, 201.
Takagi, Y., Nakayama, H., Okubo, S., Shimada, K., and Sekigachi, M. (1968). *Cold Spring Harbor Symp. Quant. Biol.* **33**, 21.
Takebe, H., and Jagger, J. (1969). *J. Bacteriol.* **98**, 677.
van de Putte, P., Westenbroeck, C., and Rorsch, A. (1963). *Biochim. Biophys. Acta* **76**, 247.
Wacker, A., Dellweg, H., and Weinblum, D. (1960). *Naturwissenschaften* **47**, 477.
Ward, H. M. (1894). *Phil. Trans. Roy. Soc. London* **185**, Part II, 961.
Weiss, B., and Richardson, C. C. (1967). *Proc. Nat. Acad. Sci. U. S.* **57**, 1021.
Witkin, E. M. (1946). *Proc. Nat. Acad. Sci. U. S.* **32**, 59.

Witkin, E. M. (1951). *Cold Spring Harbor Symp. Quant. Biol.* **16,** 357.
Witkin, E. M. (1963). *Proc. Nat. Acad. Sci. U. S.* **50,** 425.
Witkin, E. M. (1966a). *Science* **152,** 1345.
Witkin, E. M. (1966b). *Radiat. Res., Suppl.* **6,** 30–53.
Woodside, E. E., Goucher, C. R., and Kocholaty, W. (1960). *J. Bacteriol.* **80,** 252.
Yanagita, T., Mariyama, Y., and Takebe, I. (1958). *J. Bacteriol.* **75,** 523.
Zelle, M. R., and Hollaender, A. (1955). *Radiat. Biol.* **2,** 365–430.

Chapter 4

Radiation-Induced Biochemical Changes in Protozoa

GARY L. WHITSON

I. Introduction

New discoveries and the development of new methods of investigation in radiation biology have led to an expanded interest in the effects of radiations on protozoan cells. There is now a search for certain underlying biochemical changes responsible for altered biological activity in a variety of cells exposed to different radiations. Several of these studies have already been mentioned in Chapters 1 and 3. This chapter deals primarily with radiation-induced biochemical changes in DNA of protozoan cells *in vivo* and some of the techniques used to measure these changes.

To date, most of the studies in radiation biochemistry of protozoa have been in the field of ultraviolet photobiology with emphasis on the formation of photoproducts (pyrimidine dimers), their ultimate fate, and subsequent repair of DNA. In general, and as in bacterial cells, DNA is favored as the prime target responsible for ultraviolet-induced injury in cells exposed to radiations. The literature, on the other hand, is brief on this subject and more studies are needed to correlate both quantitatively and qualitatively the formation and fate of photoproducts in DNA with different kinds of observable radiation-induced injury. For this reason, it is hoped that this chapter will encourage students in radiation biochemistry to use protozoa as tools for research in the search for basic mechanisms responsible for various radiation-induced injuries to cells.

II. Survey

A. General

Studies on the biological effects of radiations on protozoa probably have their origin around the mid-1930's. Historically these studies began about the time when X-ray machines were first beginning to be used for clinical purposes in medicine. An increase in research on radiation effects followed World War II, particularly since it became apparent that X-rays and other radiations at high doses were potentially dangerous to man.

There are three comprehensive reviews on various biological aspects of radiation effects on protozoa which require very little elaboration (Giese, 1953, 1967; Kimball, 1955). Some of the observations by early investigators have included killing, retardation of cell division, loss of motility, ability to encyst and excyst, vesiculation, and increased sensitivity to heat

TABLE I

RADIATION-INDUCED BIOCHEMICAL CHANGES IN SELECTED PROTOZOA

Effect	Organism	Radiation	Reference
Nucleic acids			
Content	*Tetrahymena pyriformis*	UV	Iverson and Giese (1957)
DNA			
Synthesis	*T. pyriformis*	UV	Shepard (1965)
Metabolism	*Euglena gracilis*	UV	Gibor (1969)
Cytoplasmic synthesis	*E. gracilis*	UV	Scher and Sagan (1962)
Repair synthesis	*T. pyriformis*	UV	Brunk and Hanawalt (1967)
Excision repair	*T. pyriformis*	UV	Brunk (1970)
Dark repair replication	*T. pyriformis*	UV	Brunk and Hanawalt (1969)
Induction of pyrimidine dimers	*T. pyriformis*	UV	Whitson *et al.* (1967)
Excision of pyrimidine dimers	*T. pyriformis*	UV	Whitson *et al.* (1968)
Photoreactivation of pyrimidine dimers	*T. pyriformis*	UV	Francis and Whitson (1969)
Photoreactivation of pyrimidine dimers	*Paramecium aurelia*	UV	B. M. Sutherland *et al.* (1967)
Pyrimidine dimers	*P. aurelia*	UV	B. M. Sutherland *et al.* (1968)
RNA			
Synthesis	*Euglena*	γ-rays	Matsuoka (1967a)
Synthesis	*Blepharisma*	UV	Giese (1968)
and protein	*Amoeba proteus* (fragments)	UV	Skreb and Bevilacqua (1962)
and protein	*A. proteus* (whole cells and fragments)	γ-rays	Skreb *et al.* (1965)
and protein	*A. proteus* (whole cells and fragments)	γ-rays	Skreb and Horvat (1966)
Protein			
Content	*Euglena*	γ-rays	Matsuoka (1967b)
Nuclease activity	*T. gelleii* W	X-rays	Eichel and Roth (1953)
Enzyme systems	*T. pyriformis*	X-rays	Roth and Eichel (1955)
Succinic dehydrogenase	*T. pyriformis* GL	UV	Sullivan and Sparks (1961)
Isocitric dehydrogenase	*T. pyriformis*	UV and X-rays	Sullivan and Ehrman (1966)
Catalase activity	*T. pyriformis*	γ-rays	Roth and Buccino (1965)
Phosphomonoesterases	*Stylonychia pustulata*	UV	Hunter (1965)

and certain chemicals in cells exposed to radiations. Later investigations included the mutagenic effects of radiation (Kimball, 1964, 1966a) and more recently developmental phenomena such as regeneration (Giese, 1967, 1968; Burchill and Rustad, 1969).

The exploration of various biochemical alterations in cells exposed to radiations began in the early 1950's. A general survey of the literature concerning biochemical parameters explored in protozoan cells is shown in Table I. As indicated in this table, the literature on this subject is not very extensive. I have included only a few references on *Euglena,* but for the most part have otherwise ignored other photosynthetic protista that are sometimes considered as protozoa by protozoologists, but more often are classified as algae by phycologists.

B. Proteins

Eichel and Roth (1953) were first to report that X-rays alter nuclease activity in *Tetrahymena,* and until then virtually nothing was known about radiation-induced structural changes in biological macromolecules. They were able to measure the activities of DNase and RNase in cells exposed to X-rays and observed that 300 and 500 kR doses did not appreciably change the activities of these enzymes *in vivo.* On the other hand, they did find that these enzymes in homogenates irradiated at 500 kR were inhibited about 50%. Cysteine added to homogenates prior to irradiation at 300 kR, however, afforded about 50% protection.

Other enzyme systems have been studied in *Tetrahymena* exposed to both ionizing and nonionizing radiations. Some of these include catalase; succinic, glutamic, and malic dehydrogenase systems; and certain metabolites of these enzymes (Roth and Eichel, 1955; Roth, 1962; Roth and Buccino, 1965), such as pyruvate, phenylalanine, lactate, acetate, and thioctic acid. In general it has been found that enzyme systems are initially depressed in cells irradiated by varying kinds and doses of radiations, but that recovery is very rapid and nearly complete. Protozoans in general are very resistant to radiation and biochemical studies of the kind just mentioned were performed in the hope of establishing a biochemical basis for this resistance. Unfortunately, none of these studies provided the information necessary to establish any biochemical basis for radiation resistance at the enzyme level.

Further studies have been performed on enzymes of oxidative metabolism in irradiated synchronized *Tetrahymena* (Sullivan and Sparks, 1961; Sullivan and Ehrman, 1966). These studies were aimed at elucidating the mechanism of interference of cell division imposed by radiation. It was found (Sullivan and Sparks, 1961) that succinic dehydrogenase activity was accelerated in *Tetrahymena* irradiated at the end of the synchronizing treatment (after 7 heat shocks) and in cells irradiated late in the process of division (150 min after the last heat shock). These irradiations were performed at times when synchronized control cells showed the lowest rates of activity for this enzyme. Sullivan and Ehrman (1966) found that both ultraviolet light and X-rays would also accelerate isocitric dehydrogenase activity in cells irradiated 60 min after the end of the synchronizing treatment. This time has been shown by Zeuthen (1964), Frankel (1962), and others to be a critical physiological transition point in synchronized *Tetrahymena* beyond which most environmental blocking agents will no longer interfere with the ensuing cell division. The results of these accelerated enzyme activities by radiation, however, have been interpreted by Sullivan and co-workers as the organism's attempt to recover from radiation damage. Needless to say, none of these studies was conclusive enough to attribute to oxidative enzymes a direct positive role in the recovery processes in these irradiated cells.

Phosphomonesterases have also been studied in the UV-irradiated ciliate *Stylonychia* (Hunter, 1965). A general decrease in activity was noticed in cells of this species with increased intensity of exposure to radiations. However, such a change in phosphatase activity has not been implicated in any sense other than a general depression of enzyme activities as a direct result of radiation.

Other studies on proteins in irradiated protozoa have included studies on changes in protein content in *Euglena* and whole cells and fragments of *Amoeba proteus*. Matsuoka (1967b) examined ^{14}C-alanine incorporation into total protein (soluble protein) on a per cell basis in log phase and stationary phase *Euglena* irradiated with γ-rays. Measurements indicated that the protein content remained unchanged for several hours following irradiation while increasing with time only in nonirradiated controls. Radiation studies in *Amoeba* have been performed on whole cells, nucleate fragments, and enucleate fragments (Skreb and Bevilacqua, 1962; Skreb *et al.*, 1965; Skreb and Horvat, 1966). These reports indicate that protein synthesis is depressed in both UV- and γ-irradiated whole amoebae and that enucleate fragments of whole cells are more sensitive to irradiations than either nucleate fragments or whole cells. These authors also suggested that RNA and protein synthesis was not entirely controlled by the nucleus but further concluded that repair of radiation damage caused to RNA and protein by γ-radiation can be partially repaired by the presence of the nucleus. The amount of protection afforded against radiation was also found to correlate directly with the amount of cytoplasm surrounding the nucleus.

C. RNA

Few biochemical studies have been carried out on the nature of RNA in irradiated protozoa. Of these studies, most deal with the effects of radiation on RNA synthesis (see Table I), and one study concerns itself with the relationship of RNA synthesis to regeneration in UV-irradiated *Blepharisma* (Giese, 1968). Skreb and co-workers (see Table I) noted that RNA synthesis was inhibited in both UV- and γ-irradiated amoebae. Again, as in the effects of radiations on protein synthesis, they reported that synthesis of RNA was depressed in increasing order from irradiated whole cells, to nucleate fragments, and to enucleated fragments. Matsuoka (1967a) found that a high dose of γ-rays (1×10^5 R) resulted in a significant decrease in RNA synthesis as well as changes in UV-absorbance in fractions on methylated albumin (MAK)-column chromatographed RNA's of *Euglena*. Such a change also indicates a qualitative difference in RNA following irradiation. Giese *et al.* (1965) reported that regeneration in the ciliate *Blepharisma* was retarded by UV-irradiation and more recently Giese (1968) reported that UV-irradiation of this organism resulted in a depression of RNA synthesis as shown by decreases in the rate of ^{14}C-uridine into RNA in irradiated regenerating cells. In the latter study, Giese (1968) reported that "such synthesis is essential for the repair processes leading to regeneration." Whitson (1965) observed similar effects of depressed RNA synthesis in regenerating *Stentor* by high concentrations of actinomycin D and arrived at the similar conclusion that renewed RNA synthesis was necessary for regeneration.

D. DNA

As in many other areas of molecular biology, DNA has probably received more attention by radiation biologists than any other class of cellular macromolecules. The focal point of attention on DNA by radiation biologists include effects of radiation on content, synthesis, breaks, structural alterations, base damage, and excision and repair of pyrimidine dimers following different kinds of radiation. Radiation studies concerning DNA in protozoa, as mentioned earlier and as shown in Table I, have been more limited in scope and for the most part involve UV-induced photoproducts (pyrimidine dimers). General studies have included content and synthesis of DNA in cells, and more specific studies were concerned with qualitative changes in DNA (formation of pyrimidine dimers), repair synthesis, and repair mechanisms in cells. Only brief mention will be made here to the more generalized studies, and the more specific biochemical studies will be dealt with in the remaining sections of this chapter.

In studying the relationship of UV susceptibility and nucleic acid content in synchronized *Tetrahymena,* Iverson and Giese (1957) concluded that there was "no correlation of UV resistance with these compounds." Shepard (1965) reported that UV induced abnormal DNA synthesis in single cells of *Tetrahymena* first isolated during cell division from small logarithmically growing cultures and then irradiated at a known position in the cell cycle. The precise position of the cells irradiated during the cell cycle was determined by using one sister cell as a control and irradiating another. He noted that cells exposed to 2000 ergs/mm^2 of UV at 254 nm at any time during the cell cycle would synthesize about three times their normal amount of DNA prior to the first division following irradiation. This excess DNA was observed later to be eliminated from irradiated cells as large macronuclear extrusion bodies and cells would return to their normal DNA content after three cell divisions.

Cytoplasmic DNA synthesis has become a subject of interest in photosynthetic cells since it is apparent that chloroplasts contain DNA (for review, see Schiff and Epstein, 1968). Lyman *et al.* (1961) were the first to discover that UV inactivates chloroplast replication in *Euglena* and that this process could be reversed by photoreactivation. Scher and Sagan (1962) followed ^{3}H-thymidine incorporation in green UV-irradiated and unirradiated *Euglena.* They reported that no cytoplasmic DNA synthesis occurred in UV-irradiated cells but did occur in control cells. Unpublished studies in this laboratory (Whitson and Royal, 1968), including a variety of methods of labeling and isolation of DNA from *Euglena,* do not demonstrate any labeled DNA (either nuclear or cytoplasmic) from cells exposed to tritiated pyrimidine precursors. It has been confirmed (Buetow, personal communication, 1970) elsewhere that *Euglena* for unknown reasons does not readily incorporate exogenous pyrimidine precursors into its DNA. For this reason any further attempts to isolate labeled pyrimidine dimers from DNA in *Euglena* has been abandoned. On the other hand, attention was drawn to the possibility of identifying photoproducts as dinucleotides or trinucleotides by labeling nucleotides using ^{32}P label (R. B. Setlow *et al.,* 1964). Setlow and co-workers were able to show that nuclease-resistant sequences in the form of dinucleotides and trinucleotides could be chromatographed from irradiated bacterial DNA. Royal *et al.* (1969) reported that labeled phosphate (^{32}P) is incorporated into DNA of *Euglena* and that enzymatic hydrolysis of isolated DNA from controls resulted in the production of 5′-mononucleotides. There was evidence in UV-irradiated DNA from *Euglena* that nuclease-resistant sequences were obtained but other labeled contaminants resulted in unclear chromatograms. Studies of this kind, therefore, are feasible but more work is needed to identify photoproducts in *Euglena* using this technique.

In another study on synthesis of DNA in *Euglena,* Gibor (1969) reported that $^{32}PO_4$ incorporation into DNA is inhibited by ultraviolet light and that photoreactivation does not result in an immediate restoration of DNA synthesis. This study, however, did not show that this DNA synthesis was associated with control of the growth and development of plastids.

III. Changes in DNA following UV-Irradiation

Both direct and indirect methods have now been used to measure changes in DNA in protozoan cells following UV-irradiation. Direct methods have involved the chromatographic isolation of pyrimidine dimers from hydrolyzates of DNA and the fate of these photoproducts during dark recovery and after photoreactivation (Whitson *et al.*, 1967, 1968; Francis and Whitson, 1969; Sutherland *et al.*, 1967, 1968). Indirect methods have also been used to measure changes in DNA (fate of pyrimidine dimers) of *Tetrahymena* following UV-irradiation and have been successfully used to study the excision repair process and photoreactivation (Brunk and Hanawalt, 1967, 1969; Brunk, 1970). This section explains these different methods in detail and improved techniques for the study of photoproducts in DNA of protozoan cells.

A. Production of Pyrimidine Dimers: A Direct Measure of Photochemical Damage

Pyrimidine dimers, as already mentioned in Chapters 1 and 3, are stable photoproducts formed by covalent bonding of adjacent pyrimidines

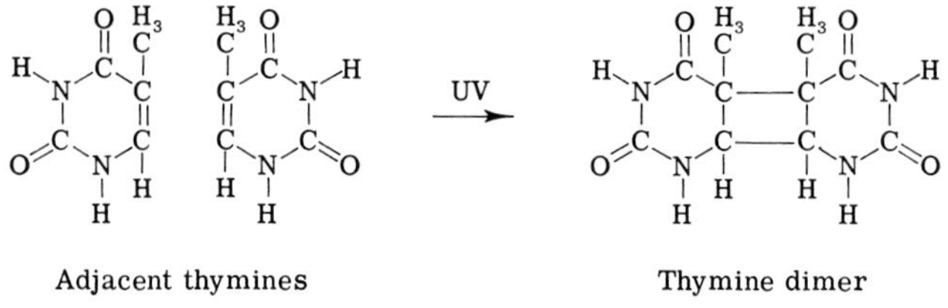

UV

Adjacent thymines

Thymine dimer

Fig. 1. The presumed formation of a thymine dimer.

in the same strand of DNA following UV irradiation. The chemical structure of a pyrimidine dimer results in the formation of a cyclobutane ring joining the adjacent bases as shown in Fig. 1. Pyrimidine dimers in DNA are resistant to acid hydrolysis and thereby can be isolated by paper chromatography on the basis of their mobilities.

1. *Growth and Labeling of Cells*

The method of culture of axenic protozoa such as *Tetrahymena* is very simple. One of many recipes consists of growing cells on 1–2% proteose peptone (Difco), 0.1–0.2% liver extract, and the following salts: NaCl (0.2% w/v), KH_2PO_4 (0.1% w/v), $Na_2HPO_4 \cdot 12H_2O$ (0.1% w/v), and $MgCL_2 \cdot 6H_2O$ (0.03% w/v). Cells are routinely transferred on a daily basis by inoculating 1 ml of cells from a day-old culture into 10 ml of fresh autoclaved medium contained in 30-ml capacity screw-capped Pyrex tubes. Cells are labeled by adding 10 μC/ml (high specific activity) ^{3}H-thymidine to a tube of cells and incubating overnight (12–15 hr) preceding irradiation. This procedure results in a good yield of organisms (25,000–50,000 cells/ml) and ensures uniform labeling of cells over several generations. The normal generation time for *T. pyriformis* GL-C is about 3–3.5 hr in tubes grown on a slant at 26°–27°C.

Organisms such as *Paramecium aurelia* maintained routinely in bacterized cultures (*Aerobacter aerogenes*) have long generation times and do not readily incorporate ^{3}H-thymidine from the medium (Berger and Kimball, 1964). Therefore, cells are usually grown on *Escherichia coli* which are prelabeled with this radioactive precursor. Feeding is allowed for several hours prior to irradiation to ensure active uptake of the label (B. M. Sutherland *et al.*, 1968).

2. *Administering UV*

Irradiation is performed either by the use of a monochromator set at a wavelength about 265 nm (Whitson *et al.*, 1968) or with a germicidal lamp (254 nm) as reported by B. M. Sutherland *et al.*, 1968. It is important in either situation to remove cells from opaque medium and irradiate them in a standard salt buffer medium. Sutherland and coworkers used Dryl's solution for paramecia, whereas, in our studies with *Tetrahymena,* we use a salt solution consisting of the salts in the same concentration as in the media mentioned above for growth of cells. In UV experiments, the concentration of cells is also an important factor. B. M. Sutherland *et al.* (1968) found that there was negligible shielding of cells from UV in nonstirred samples in 10 ml of Dryl's solution in a 6.0-cm petri dish if the cell population did not exceed 10^4 cells/ml. We have observed that stirred samples irradiated in 1-cm quartz cuvettes are optimal if the cell concentration in 2 ml of salt solution does not exceed 10^3 cells/ml.

It is well known that thymine-containing pyrimidine dimers constitute only a small proportion of the labeled thymine in DNA and are not easily detected in protozoan cells at doses much lower than 1000 ergs/mm^2. Therefore, we have used only incident doses from 1000 ergs/mm^2 to 10,000 ergs/mm^2. *Tetrahymena* will survive for at least 24 hr after an incident dose of 2000 ergs/mm^2. Single cell studies, however, indicate that only 2% survival is observed in clones after this time (Whitson and Francis, 1968a). Paramecia are apparently less sensitive to UV and show 100% survival at 48 hr following an incident dose of 4000 ergs/mm^2 (B. M. Sutherland *et al.*, 1968).

3. *Hydrolysis of DNA*

Pyrimidine dimers are detected in DNA hydrolyzates by chromatographic isolation. Carrier and Setlow (1971) have recently published their methods in detail for detection of pyrimidine dimers in a variety of cell types. In some instances they indicate that identification of photoproducts is enhanced when DNA is first removed from lysates by extraction procedures such as those described by Marmur (1961) and in other instances by hydrolysis of whole cells and subsequent chromatographic isolation. In our studies we find that both methods are suitable for the detection of dimers. Hydrolysis of samples of extracted DNA is accomplished with 97% formic acid in sealed tubes for 1 hr at 175°C (Whitson *et al.*, 1968). In the case of hydrolysis of whole cell lysates, we use 100% triflouroacetic acid (TFA) for 1 hr at 155°C. As indicated by Carrier and Setlow (1971), formic acid is preferred because it results in cleaner chromatograms. On the other hand, TFA produces a better hydrolysis when samples contain a lot of cellular material (i.e., large protozoan cells).

4. *Chromatography*

Chromatography of formic acid hydrolyzates is performed in the manner shown in Fig. 2. Two-dimension paper chromatography is performed with Whatman No. 1 filter paper in order to get good separation of pyrimidine dimers from the free bases of the hydrolyzed DNA. The first dimension of the chromatogram is run in *N*-butanol–acetic acid–water (80:12:30) as described by Smith (1963) and the second dimension is developed in *N*-butanol–water (86:14) using the method of Wacker *et al.*

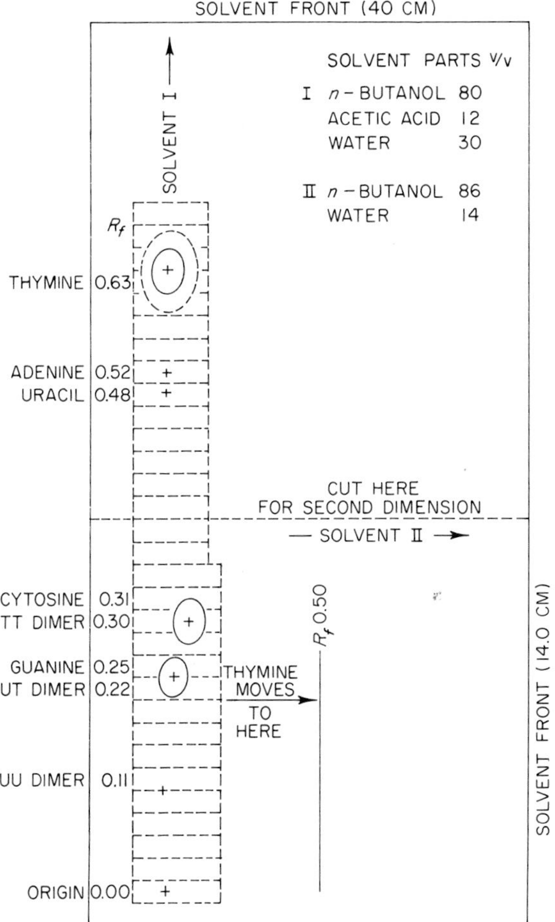

Fig. 2. A schematic representation showing two-dimensional paper chromatography for the separation of pyrimidine dimers on Whatman No. 1 filter paper. Cut along the dotted lines for the 1-cm strips. (Courtesy of W. L. Carrier.)

(1960). One-centimeter strips of the chromatogram are placed into scintillation vials and eluted with 1 ml of water. The eluates containing the strips are then mixed with 10 ml of dioxane-2,5-bis-2(5-*tert*-butylbenzoxazolyl)-thiophene (BBOT)-naphthalene scintillator [BBOT, 12 gm; naphthalene, 500 g; and dioxane (Eastman), 4 l]. The counts in the dimer region (^{3}H) are plotted as a function of position on the chromatogram and produce a radioactive profile such as shown in Fig. 3. $\widehat{UT}$ shown in Fig. 3 has been shown to result from the conversion of cytosine to uracil by deamination during acid hydrolysis (R. B. Setlow and Carrier, 1966).

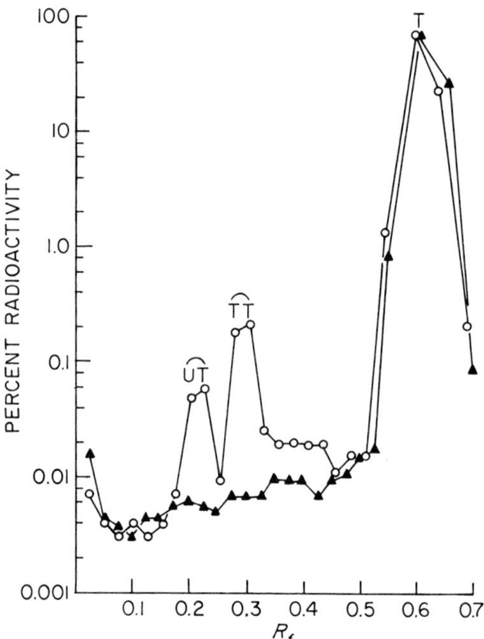

Fig. 3. The distribution of radioactivity along chromatograms of acid hydrolyzates of *Tetrahymena* DNA labeled with ^{3}H-thymine. Both ultraviolet-irradiated (5400 ergs/mm^2, $\lambda = 265$ nm) (○—○) and nonirradiated (△—△) cells are shown. $\widehat{UT}$ represents a cytosine–thymine dimer (see text); $\widehat{TT}$, a thymine dimer; and T, thymine. (From Whitson *et al.*, 1968; courtesy of the publisher.)

5. *The Excision Process*

Several different models have been proposed for the excision of thymine dimers during repair of irradiated DNA *in vivo*. The most widely held model for this excision process is based mainly on the studies of R. B. Setlow and Carrier (1964) and Kelly *et al.* (1969) on the repair of *E. coli* DNA. A scheme illustrating this process is shown in Fig. 4 and is described as follows. When pyrimidine dimers are produced in cells or DNA *in vitro* irradiated with UV, the first result is the linking together of adjacent pyrimidine residues in the same strand by the cyclobutane ring as shown in Fig. 1. This results in a distortion of the DNA helix as shown in Fig. 4A (Marmur *et al.*, 1961; Pearson and Johns, 1966). According to R. B. Setlow *et al.* (1969) an endonuclease specific for UV-irradiated DNA recognizes this distortion and makes an incision (single-strand break) near the dimer (Fig. 4B). The next step in the repair of UV-irradiated DNA is the excision of the damaged region (Fig. 4C). It ap-

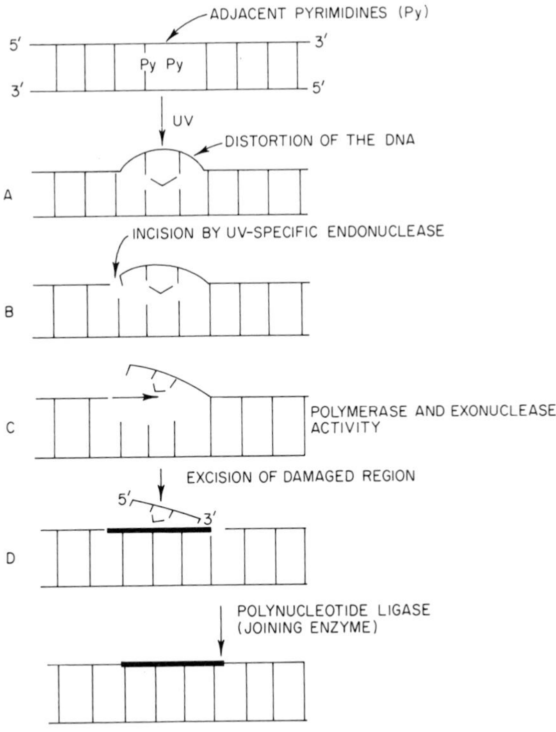

Fig. 4. A model for the excision repair of UV-irradiated DNA.

pears almost certain that this excision process is also enzymatic, but conflicting reports on the nature of this enzyme leave some doubt as to whether it is a UV-specific exonuclease or whether this function is mediated by enzymes that serve multiple functions. For instance, Kelly *et al.* (1969) have reported that DNA polymerase from *E. coli* also contains exonuclease activity that endows it with the capacity to excise damaged regions in DNA and at the same time directs the synthesis of new DNA (Fig. 4, C and D). This finding led to an editorial comment in *Nature (London)* [**224,** 412 (1969)] posing the question as to whether DNA polymerase is for repair or replication of DNA. To add to this controversy, De Lucia and Cairns (1969) have isolated a mutant of *E. coli,* strain P3478, that is deficient in DNA polymerase. This mutant is capable of normal DNA replication, excises pyrimidine dimers (Boyle *et al.,* 1970), and performs delayed but normal repair replication (Hanawalt *et al.,* 1970).

The last step in the repair of DNA is open to less question. It appears certain that the joining of the newly repaired region to the remaining region in the DNA strand is performed by a polynucleotide-joining enzyme (ligase) as shown in Fig. 4E. This enzyme apparently intervenes polymerase only after it has replaced the entire damaged region in the DNA (Kelly *et al.*, 1969).

The excision of pyrimidine dimers from DNA has been reported to occur in both *Paramecium* (B. M. Sutherland *et al.*, 1968) and in *Tetrahymena* (Whitson *et al.*, 1968). In both instances, the time required for excision takes several hours (20–25 hr for *Paramecium* and up to 50 hr for *Tetrahymena*).

It is imperative that UV-irradiated cells be kept in the dark or under a yellow light during the period of sampling to measure excision. This is necessary in order to prevent photoreactivation which will be discussed in detail in the following section. Results of a typical experiment examining the fate of dimers in UV-irradiated paramecia are shown in Fig. 5. As indicated, there appears to be a delay of about 6 hr after irradiation before dimers begin to disappear from the trichloroacetic acid (TCA)-insoluble fraction. After this time, a loss of dimers from the TCA-insoluble fraction and an increase in radioactivity in the incubation medium occurs. The excision process in *Tetrahymena* apparently is not delayed following irradiation and the loss of radioactivity from the TCA-

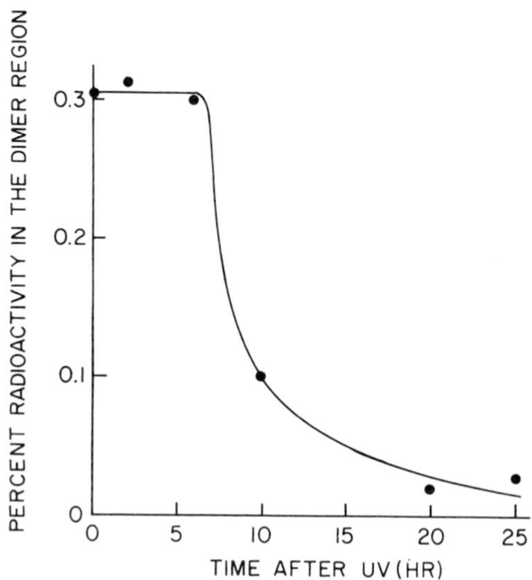

Fig. 5. A typical experiment that shows the disappearance of dimers from the TCA-insoluble fraction of UV-irradiated paramecia incubated in the dark. The dimers do not appear in the TCA-soluble fraction. (From B. M. Sutherland *et al.*, 1968; courtesy of the authors and the publisher.)

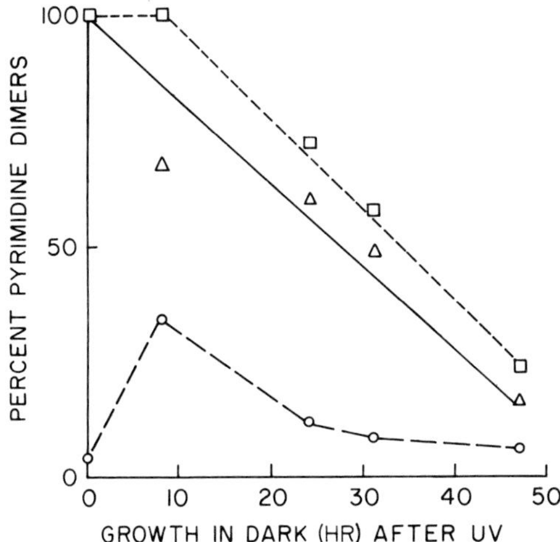

Fig. 6. The percent total thymine-containing pyrimidine dimers observed in the TCA-insoluble fraction (△—△), TCA-soluble fraction (○- -○), and both fractions combined (□ - - - □) after postirradiation growth of cells in the dark at varying periods of time. UV-irradiation: 2520 ergs/mm^2 at 265 nm. (From Whitson *et al.*, 1968; courtesy of the publisher.)

insoluble fraction proceeds more or less linearly with time (Whitson *et al.*, 1968). A typical experiment is shown in Fig. 6. Loss of radioactivity from both the TCA-insoluble and eventually the TCA-soluble fraction indicates that radioactivity is being excreted by cells into the medium.

6. *Photoreactivation*

The first demonstration of *in vivo* photoreactivation (monomerization) of pyrimidine dimers in a eukaryotic cell was reported from studies on *Paramecium* by B. M. Sutherland *et al.* (1967). However, biological photoreactivation (photoreversal) of UV damage was reported several years earlier (Kimball and Gaither, 1951; Christensen and Giese, 1956). The discovery of photoreactivation (photoreversal) of UV damage was made by Kelner (1949) and the subsequent discovery of an enzymatic basis for this process by Rupert (1960). More recent studies have shown that photoreactivating enzyme in the presence of near-UV light splits pyrimidine dimers *in situ* in DNA (Rupert, 1962; Wulff and Rupert, 1962; J. K. Setlow and Setlow, 1963; Cook, 1967, 1970).

Photoreactivation requires the administration of near-UV or visible light following irradiation of cells with far-UV to split dimers (monomerize) in DNA. Several light sources are commercially available that are suitable for this type of experiment. Some light sources are mentioned by

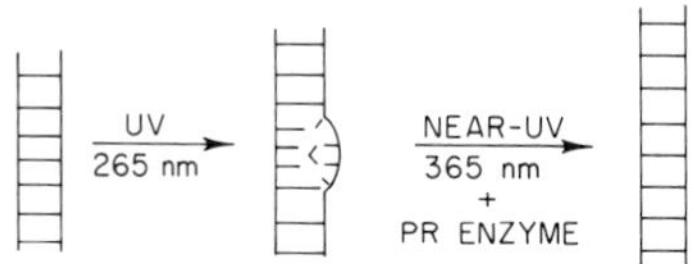

Fig. 7. The generalized scheme for monomerization of pyrimidine dimers.

Rahn (Chapter 1), Hamkalo (Chapter 3), and by Jagger (1967). The general pattern of monomerization follows according to the scheme shown in Fig. 7. The main difference between photoreactivation and dark repair is that the pyrimidine dimers are not excised but instead are destroyed

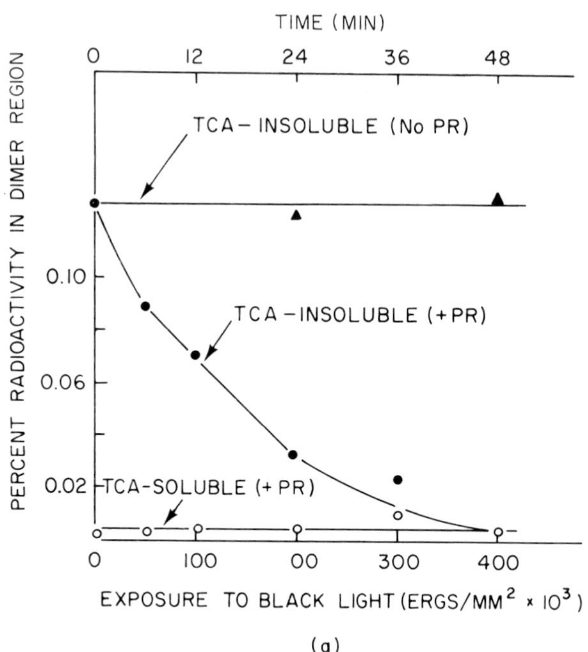

Fig. 8. a. The thymine-containing dimers in the TCA-insoluble (●) and TCA-soluble (○) fractions of paramecia irradiated with ultraviolet light after several exposures to photoreactivating light (PR). As a control, irradiated animals were kept in the dark after UV-irradiation for the same periods of time as were used for photoreactivation. The dimers in the TCA-insoluble fractions (▲) were used as controls. (From B. M. Sutherland *et al.*, 1967; courtesy of the authors and the publishers.) **b.** Loss of thymine-containing pyrimidine dimers in *Tetrahymena* during illumination with black light. Dimers were induced by irradiation with 1200 ergs/mm² at 265 nm. No decrease in dimers was seen during 110 min incubation in the dark, and no dimers appeared in the incubation media during photoreactivation treatment. (From Francis and Whitson, 1969; courtesy of the authors and the publisher.)

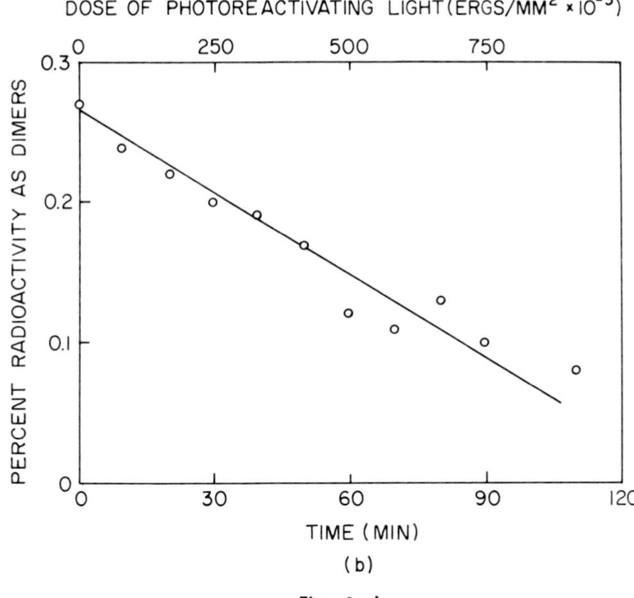

Fig. 8. b.

in situ by removal of the cyclobutane ring joining adjacent pyrimidines. Thus when hydrolyzates of DNA are examined after photoreactivation (PR) there is a loss of dimers from the TCA-insoluble fraction but no accumulation of radioactivity (dimers) in the TCA-soluble fraction.

The monomerization of pyrimidine dimers by black-light photoreactivation has been examined in both *Paramecium* (B. M. Sutherland *et al.*, 1967) and in *Tetrahymena* (Francis and Whitson, 1969). In both instances, a loss of radioactivity was observed in the dimer regions of chromatograms after this treatment. To determine the time required for completion of photoreactivation, samples of UV-irradiated cells were removed periodically during continuous exposure to photoreactivating light. Typical results for paramecia are shown in Fig. 8a and for *Tetrahymena* in Fig. 8b.

The rate of monomerization of dimers in *Tetrahymena* appears to be linear with time and dependent upon the dose rate. This may be an indication that the amount of photoreactivating (PR) enzyme available for complexing with dimers during the photoreactivating process is limiting (Jagger, 1967). It is certain, however, that the rate of monomerization, at least in *Tetrahymena,* is dependent on both the wavelength and intensity of monochromatic photoreactivating light as well as the temperature of the cell sample (Francis and Whitson, 1969). An action

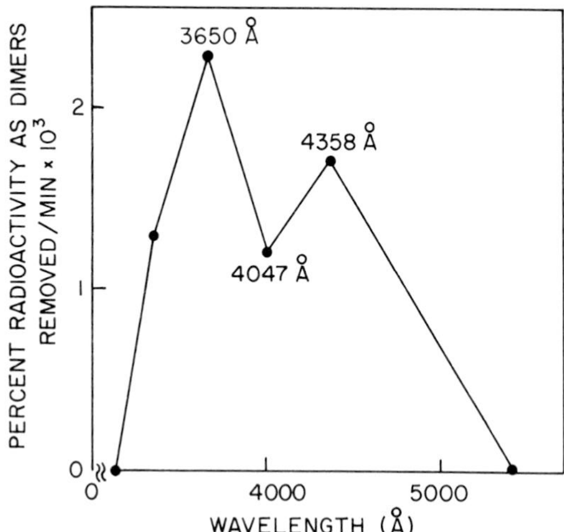

Fig. 9. Relative effectiveness of different wavelengths of monochromatic photoreactivating light in the monomerization of thymine-containing pyrimidine dimers *in vivo* in *Tetrahymena*. Each point represents the averaged slopes of at least two dose curves at the given wavelength. In each case the intensity of photoreactivating light was 6000 ergs/mm^2/min. The ultraviolet (265 nm) dose in each case was 1200 ergs/mm^2. (From Francis and Whitson, 1969; courtesy of the authors and the publisher.)

spectrum of PR obtained with a Bausch and Lomb Monochromator using lines from the mercury arc source adjusted to an intensity of 6000 ergs/mm^2/min is shown in Fig. 9. This action spectrum is somewhat different from that of bacteria and other organisms. According to Jagger (1967) this may indicate the probability of "more than one enzyme—and more than one chromaphore" involved in photoreactivation. The two peaks of relative effectiveness of the action spectrum for PR shown in Fig. 9 (365 and 436 nm) support Jagger's contention.

B. Repair Replication as a Measure of Photochemical Damage

Repair replication of DNA is another way by which photochemical damage can be assayed in organisms. Repair replication is termed "nonconservative" DNA replication because it differs from the normal "semiconservative" DNA replication in that portions of damaged parental DNA strands are excised (pyrimidine dimers in the case of photochemical damage) and replaced rather than conserved (Brunk and Hanawalt, 1967). Repair replication also differs from normal DNA replication in that after this replication there is no net increase in the amount of DNA present. Physical evidence for this mode of replication was first observed in bacterial cells by Pettijohn and Hanawalt (1964). Their technique involved the use of double-isotope labeling (both tritium and ^{32}P), a heavy density label (5-deoxybromouridine), and analysis of DNA in both neutral and alkaline CsCl gradients. An improved method of the study of repair replication as a measure of dark repair (excision and replacement of dimers) and photoreactivation (monomerization of dimers) in protozoa is the main concern of the ensuing sections.

1. *Method of Study*

a. Growth, Labeling, and Irradiation of Cells. The methods presented here are adapted from those of Brunk and Hanawalt (1969) on repair replication in UV-irradiated *Tetrahymena.* Cells can be grown in either defined medium as described by Elliott *et al.* (1954) supplemented with 0.04% proteose peptone (Difco) or in 1–2% proteose peptone with 0.1–0.4% liver extract and salts as described in Section III,A,1. *Tetrahymena* has a long generation time on defined medium, sometimes 2–3 times longer than on the latter mentioned, more crude medium that produces a generation time of 2–3 hr. Labeling of cells is performed by growing populations for several generations in the presence of ^{3}H-thymidine. The amount of label required depends on the specific activity of the compound. Brunk and Hanawalt (1969) used 0.5 μC/ml ^{3}H-thymidine, specific activity 11.4 Ci/mmole. Whitson *et al.* (1968) found, by labeling cells with 10 μC/ml of low specific activity (3.0 Ci/mmole) ^{3}H-thymidine, that there was no loss of label in total DNA or damage to control cells in culture for up to 50 hr following this treatment.

This technique for assaying excision and replacement of damaged regions of DNA is more sensitive than the chromatographic detection of pyrimidine dimers and therefore allows one to employ smaller doses of UV. Brunk and Hanawalt (1967, 1969) were able to make measurements in cells after exposures of only 250 ergs/mm^2 of UV from a 32-W germicidal lamp. Immediately after irradiation, cells are placed in culture containing the high density label (10^{-50} μg/ml 5-bromodeoxyuridine) and 10^{-40} μC/ml of ^{32}P-phosphoric acid and allowed to grow for one full generation.

b. DNA Isolation. DNA is isolated from lysed cells washed in a suitable buffer. Brunk and Hanawalt (1969) used NET buffer (0.5 *M* NaCl, 0.05 *M* EDTA, 0.05 *M* Tris, pH 9.5). Lysing is performed after pelleting cells by low-speed centrifugation (500 *g*) by treatment of washed cells for 0.5 hr with pronase (100 μg/ml) and Sarkosyl-NL-97 (Geigy) (0.1%). Lysates are extracted with an equal volume of chloroform–octanol (9:1) to remove the bulk of protein. Lysis of cells can also be carried out in sodium lauryl sulfate ($\sim$0.5%) at room temperature for 30 min (Whitson and Francis, 1968a) and then precipitation of the protein by chloroform–isoamyl alcohol (24:1) (Whitson *et al.*, 1968). Highly purified DNA can be obtained by further extraction techniques (Marmur, 1961), but these are not advisable or desirable because DNA is often fragmented during prolonged isolation procedures and furthermore are not necessary for isolation on CsCl gradients.

c. Density Gradient Centrifugation. Centrifugation of DNA is performed in the heavy salt cesium chloride by ultracentrifugation for several hours. It has been observed by this technique (Meselson and Stahl, 1958) that a solution of CsCl ($\sim$1.70 gm/cm^3) spun at $\sim$37,000 rpm will form its own gradient varying from about 1.5 gm/cm^3 at the top of the tube to about 2.0 gm/cm^3 at the bottom of the tube. *Escherichia coli* DNA in such a gradient has a mean density of 1.71 gm/cm^3. DNA from *Tetrahymena* has two peaks of average density in CsCl: a major peak at 1.688 gm/cm^3 and a minor peak at $\sim$1.684 gm/cm^3 (Suyama and Preer, 1965). Strain differences of this species may result in slight differences in these values (Whitson *et al.*, 1969). Presumably the major peak constitutes the bulk of macronuclear DNA while the smaller, less-dense peak is mitochondrial DNA. Brunk and Hanawalt (1967, 1969) reported only relative values of bulk DNA for *Tetrahymena* and record only drop number as will be described later.

Parental and daughter strands of DNA can be separated the one from the other by rebanding DNA in alkaline CsCl adjusted to pH 12.9 with 0.5 *M* K_2HPO_4 and KOH (Brunk and Hanawalt, 1969).

The method of study just outlined is unique in that it allows one to (1) separate hybrid DNA from normal DNA; (2) separate newly synthesized daughter strands from parental strands of DNA; and (3) to use it as an indirect measure of the repair of UV-irradiated DNA. All of these possibilities arise from the fact that 5-bromouracil (BUdR) is an analog of thymidine and is readily incorporated into DNA in place of the natural base thymine. The bromine atom of BUdR is five times heavier than the methyl group in normal thymine; therefore, DNA fragments that contain this analog are heavier than normal DNA fragments and thus can be separated from them in CsCl gradients.

2. *Measurement of Dark Repair*

It can be said now that repair replication is an indirect measure of the sum of all of the excision–repair processes in DNA (dark repair) after ultraviolet-induced injury to cells. The results of excision–repair in *Tetrahymena* are shown in Fig. 10. The radioactivity profiles shown are

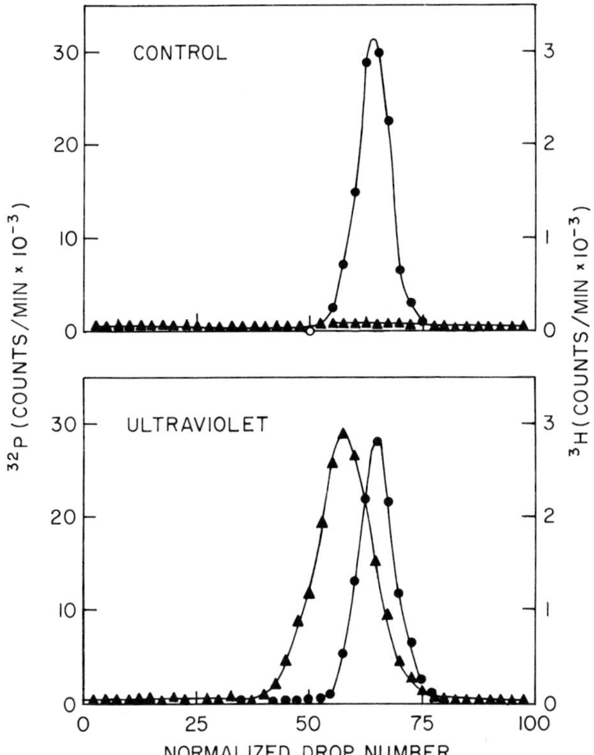

Fig. 10. Alkaline CsCl density gradient analysis of parental DNA strands that have undergone excision repair. An exponentially growing culture of *Tetrahymena pyriformis* was labeled with ^{3}H-TdR before UV-irradiation and with ^{32}P after irradiation. The BUdR-labeled hybrid density band of DNA from a neutral CsCl density gradient was isolated and then rerun in the alkaline gradient to separate daughter strands (BUdR-labeled) from parental strands (thymine-labeled). The isolated parental material was then rerun in a second alkaline CsCl gradient as illustrated above. Upper figure, unirradiated control; lower figure, UV-irradiated culture. ●, ^{3}H label; ▲, ^{32}P label. (From Brunk and Hanawalt, 1969; courtesy of the authors and the publisher.)

only those of isolated parental DNA strands after separation from hybrid DNA fragments and then rerun in the alkaline gradient. The presence of ^{32}P in parental strands in UV-irradiated cells suggests that all steps of the excision–repair process have been completed. Note that there was no incorporation of ^{32}P in the control culture.

3. *Reversal of Damage in Situ*

As mentioned previously in Section III,A,5, photoreactivation involves the reversal of photochemical damage (monomerization of pyrimidine

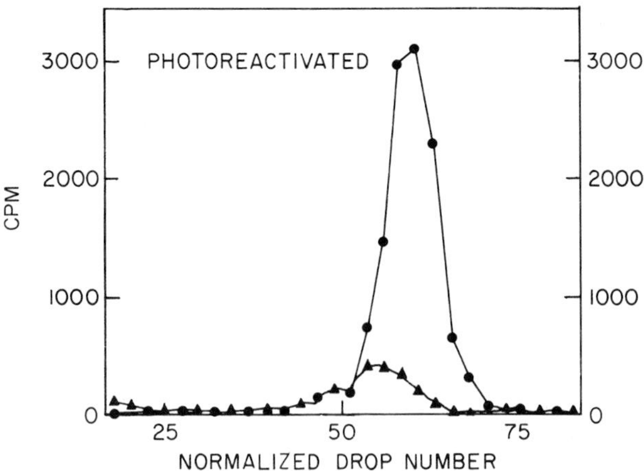

Fig. 11. Fractions from an alkaline CsCl gradient containing parental DNA strands after photoreactivation. ●, ^3H label before radiation; ▲, ^{32}P label after irradiation. (From Brunk and Hanawalt, 1967. Courtesy of the authors and the publisher.)

dimers) *in situ*. Such a reversal does not involve the excision of labeled photoproducts from cells. Therefore, one can expect that when UV-irradiated cells prelabeled with ^3H-thymidine are subjected to photoreactivating light in the presence of 5-bromouracil and another label (^{32}P) that there would be no incorporation of these latter compounds. Thus there would be no net change in the radioactivity or an increase in the density of some of the DNA fragments. A typical result is shown in Fig. 11. A small amount of label (^{32}P) appears to have been incorporated, however, possibly indicating that complete PR was not obtained in the sample examined and possibly also an indication that a small amount of excision–repair had taken place. Nevertheless, a comparison of the results shown in Fig. 10 with those in Fig. 11 clearly indicates the difference of these two repair processes.

IV. Biochemical Effects of Ionizing Radiation

Although evidence has been obtained from studies on bacterial and mammalian cell systems that favor the idea that DNA is an important if not the most prime target involved in lethal and sublethal ionizing radiation (Kanazir, 1969), literature on such studies in protozoa is almost entirely lacking. Giese (1967) suggests that DNA is suspect "since it is known to control synthesis of enzymes and to regulate their normal functioning." There are some data from studies with *Tetrahymena* that favor DNA as an important target for both sublethal and lethal ionizing radiations. For instance, Brunk and Hanawalt (1967) discovered that 80,000 rads of X-rays (a sublethal dose) promotes about the same amount of repair replication as 250 ergs/mm^2 of UV and that this change is not photoreactivable. In unpublished results we have observed that 60,000 rads of X-rays delay synchronized cell division in *Tetrahymena* for nearly 3 hr. Analysis of irradiated cells on alkaline sucrose density gradients indicate that single-strand breaks in DNA result from irradiation and that these breaks are dark-repaired during the 3 hr preceding cell division. Other data obtained in this laboratory (Mossman and Whitson, 1971) indicate that large doses of ionizing radiations (700 kR of γ-rays from a cobalt-60 source) combined with actinomycin D (15 μg/ml) kills more than 50% of the cells and acts synergistically to induce the breakdown of DNA in *Tetrahymena.* The latter findings are only preliminary observations and most of the evidence implicating DNA as primarily responsible for the lethal effects of ionizing radiation comes from studies on radiation-induced breakdown of DNA in bacteria (Stuy, 1960, 1961; Pollard and Achey, 1964; McGrath *et al.*, 1966; Grady and Pollard, 1967).

There is a controversy in the literature concerning whether actinomycin D protects DNA from or sensitizes DNA to ionizing radiation. Kollman *et al.* (1969) observed a protective action of salmon-sperm DNA by actinomycin, whereas Grady and Pollard (1967) observed a sensitizing effect of bacterial DNA to ionizing radiation by actinomycin. In this laboratory (Mossman and Whitson, 1971) we find that low doses of actinomycin D (2 μg/ml) radioprotect cells, whereas, high doses of actinomycin D (15 μg/ml) radiosensitize *Tetrahymena* DNA *in vivo.* According to Grady and Pollard (1967) there are two possible mechanisms for the sensitization of bacterial cell DNA by actinomycin to ionizing radiation. These are: (1) Actinomycin D inhibits either existing DNA repair systems or the synthesis of any repair enzyme that would be induced by the action of radiation; or (2) actinomycin D binds to DNA in such a way as to render degradation more rapid and effective. This, however, does not solve the existing controversy, and, furthermore, Grady and Pollard (1967) hold the view that degradation of DNA is enzymatically mediated following exposure of cells to lethal doses of ionizing radiation. Yet definite proof that this degradation is mediated by enzymes is still lacking. There are reasons to suspect, however, that enzymes may be involved in this process. One of the reasons is supported by the earlier findings of Roth and Eichel (1955) that large doses of X-rays (500 kR) do not appreciably alter the activity of DNase in *Tetrahymena.*

Most studies concerning ionizing radiation effects on proteins in protozoa have involved studies on enzyme activities (see Table I). There is evidence, again from bacterial and mammalian cell studies, that radioresistance is somewhat dependent on the sulfhydryl (SH) content of cells (Alexander *et al.*, 1965; Bruce and Malchman, 1965; Revesz *et al.*, 1963; Revesz and Modig, 1966). The conclusion reached was that cells containing higher amounts of SH compounds are in general more resistant to ionizing radiations. That is apparently due to the fact that SH groups are free radical scavengers and that the "protection effect" afforded by compounds containing large amounts of SH is due to the ability of SH groups to absorb radicals produced by ionizing radiations. This has led to the formulation of the "sulfhydryl hypothesis" that radiation first damages SH groups before affecting other cellular constituents (for a review, see Okada, 1969). It is also known that some of the SH groups in cells are contained in part in cellular proteins (to be discussed in more detail later on) and appear to be the active sites of some enzymes. It is entirely possible, therefore, that altered enzyme activities following irradiation of protozoan cells may be a direct result of such action.

It is generally known that protozoans have a natural resistance to large doses of ionizing radiation. Some of this resistance may be attributed to SH groups in these cells. It is more probable, however, that a large part of the resistance is due to the amount of ploidy of DNA in these organisms. Most ciliates, for instance, are highly polyploid and contain up to several hundreds of sets of genomes. Data obtained from studies on mammalian cells (Storrer, 1967) support the idea that radioresistance of cells increases with increased ploidy. It may also be possible that the resistance of many protozoans to ionizing radiations is due in part to highly efficient DNA repair mechanisms. It is known that certain radiation-resistant strains of bacteria have efficient repair mechanisms (Moseley and Laser, 1965; Dean *et al.*, 1966; McGrath *et al.*, 1966).

The resistance of protozoa to ionizing radiations may vary at certain times during the cell cycle. Some evidence with paramecia indicates that cells are more sensitive to ionizing radiations during DNA synthesis than at other times during the cell cycle (Kimball, 1966b). This supports the idea that errors in DNA replication are involved in radiation-induced mutagenesis and that if enough time is allowed for DNA repair before actual synthesis takes place, then lethal or other mutagenic effects can be reversed. Actual physical and chemical measurements of DNA in radiation-induced mutant protozoans have not been examined. Investigations of this kind are difficult, if not impossible, to perform with protozoa because auxotrophic mutants such as in bacteria are rarely if ever induced by radiations. The only observable marker for radiation-induced mutation up to this time has been cell deaths. Clones of cells destined to die following lethal ionizing radiation and after autogamy (a process of self-fertilization in ciliates resulting in all organisms being homozygous) yield too small of an amount of material for the usual biochemical analyses necessary to determine the nature of the errors in DNA.

There are some indications (see Section II,C) that RNA is an important target for ionizing radiations in protozoan cells. But it is not entirely clear whether lethal effects or long-lasting effects of ionizing radiation are due to damage to RNA in cells. Giese and Lusignon (1960), from studies with *Blepharisma,* suggest that X-ray damage to RNA blocks cell division. Their findings indicate that recovery is more rapid in irradiated cells that are cut and allowed to regenerate than in cells irradiated and left uncut. Their basic assumption is that renewed RNA synthesis is necessary for regeneration. If irradiated cells are cut then large portions of damaged RNA are removed and replaced by normal RNA. This in turn allows normal regeneration and subsequent cell division to occur. The nature of the radiation-induced damage to RNA is unknown, but it most certainly affects the synthesis of proteins involved in normal cell processes.

V. Extensions

More work is needed to extend our knowledge concerning radiation biochemical effects of protozoa. The following list includes some recommendations of the direction this work might take.

A. Correlation of Biological Effects with the Formation and Fate of Photoproducts

One area that is unexplored concerns the correlation of both quantitative and qualitative changes in photoproducts with the inhibitory effects (prevention of cell division) and the lethal effects produced by ultraviolet irradiation. Preliminary investigations with *Tetrahymena* have been aimed at part of this problem (Whitson and Francis, 1968b; Whitson, 1969). It was observed that an action spectrum for killing did not correlate quantitatively with the *in vivo* formation of pyrimidine dimers alone. That is to say, it was observed that twice as much killing in *Tetrahymena* resulted in single isolates of cells irradiated at 280 nm than at 265 nm. The same incident doses at both of these wavelengths, however, produced the same number of chromatographically isolatable thymine-containing pyrimidine dimers. It is not known whether cytosine–cytosine dimers are more prevalent after irradiation at 280 nm or if certain hydrates could be responsible for increased killing. Data by Rahn (Chapter 1) certainly point to the fact that cytosine hydrates as well as other possible photoproducts could be responsible for some of the cell killing produced by UV.

B. The Role of SH-Containing Proteins

Action spectra on ciliary reversal, immobilization (Giese, 1945a), and the retardation of cell division (Giese, 1945b; Kimball *et al.*, 1952) with a peak at 280 nm indicate that damage to proteins might be at least partially responsible for increased killing in cells exposed to ultraviolet light. It has also been argued that proteins that are rich in SH groups play an important role in cell division (Dan, 1966). Watanabe and Ikeda (1965) have gone so far as to claim that they have isolated a "division protein" that is rich in SH content from *Tetrahymena*. This protein also readily incorporates the sulfur-containing amino acid methionine. According to the "sulfhydryl hypothesis," absorption of this protein with ionizing radiations could interfere with cell division or even result in cell death, that is, if this protein is indeed a division-directing protein. Therefore a combined study on the effects of ionizing radiation on cell division and change in this protein might lead to an answer to this important problem.

C. Radiation Effects on Membranes

Another hypothesis favors the idea that radiation damage to cell membranes is responsible for much of cell killing (Bacq and Alexander, 1961). However, very few if any such implications have been made concerning protozoan cell death. On the other hand, studies on *Amoeba* indicate that pinocytosis is induced by ultraviolet light (Rinaldi, 1959). No doubt stimulation of this process is related to membrane changes. It is known, however, that the release of enzymes from membrane-bound subcellular organelles such as mitochondria (Goldfeder, 1963), nuclei (Hagen *et al.*, 1963), and lysosomes (Goutier-Pirotte and Thonnard, 1956) results from ionizing irradiation of certain mammalian cells.

Studies on red blood cells indicate that there may be an interaction of sulfhydryl compounds with membrane changes that may be responsible for radiation damage to cells. For instance, there is some evidence that increased permeability of red blood cells involves SH groups (Shapiro *et al.*, 1966; R. M. Sutherland *et al.*, 1967). This evidence comes from the fact that (1) SH-blocking agents such as *p*-hydroxymercuribenzoate result in similar increased K^+ leakage and increased Na^+ uptake as does ionizing radiation, and (2) the SH compound mercaptoethylguanidine protects cells from radiation-induced uptake of Na^+ (Shapiro *et al.*, 1966). Thus, it could be that the killing and inhibitory effects of ionizing radiations of protozoa may involve both membranes and SH-containing proteins. The large size, availability, and the ease of culturing and handling of protozoans make them suitable material for investigations of this kind.

D. Repair Mechanisms

Although repair mechanisms for damage produced in DNA by nonionizing (UV) radiations are apparently well understood, mechanisms for repair of DNA in cells exposed to ionizing radiations remain a mystery. This is particularly the case with protozoan cells. Perhaps future ventures in radiation biology with protozoa will lead us to a clearer understanding of these and other basic mechanisms involved in radiation-induced damage to cells.

VI. Summary

Studies on the radiation biochemistry of the protozoa to date have included research on DNA, RNA, and proteins. *In vivo* experiments with DNA have dealt primarily with the effects of nonionizing radiations (UV) on content, synthesis, and production and fate of pyrimidine dimers in DNA. Studies on both the chromatographic isolation of pyrimidine dimers and the repair replication of DNA in certain protozoan cells indicate that there are efficient repair mechanisms involved in dark repair and photoreactivation. There have been a few studies on RNA in certain protozoan cells following irradiation, but these have dealt primarily with content and synthesis rather than any alteration in structure. The main discovery that has come out of RNA studies is that RNA is necessary for cell regeneration and cell division. Less is known concerning the effects of radiation on proteins. Most of these studies, which have been done mainly on enzyme activities, indicate that proteins *in vivo* are in general very resistant to ionizing radiations. Very few relationships have been established concerning direct biochemical changes in macromolecules as a basis for alteration of biological activities, e.g., ciliary immobilization, prevention and delays of cell division, inhibition of regeneration, and lethality. Findings from other cellular studies in radiation biochemistry suggest the possibility that changes in SH compounds and alteration of cell membranes may be important targets for ionizing radiations in protozoan cells.

Acknowledgments

Previously unpublished work was supported by NIH Biomedical Sciences Support Grants 141080 1173R 01 and 121080 1173R 02 to the University of Tennessee and a Faculty Research Fund Grant from the University of Tennessee. The author expresses his thanks to Miss Anita Walker and Miss Sarah Fritts for help in preparing this manuscript. The author also expresses his thanks to Dr. K. W. Jeon for critical review of the manuscript.

References

Alexander, P., Dean, C. J., Hamilton, L. D. G., Lett, J. T., and Parkins, G. (1965). *In* "Cellular Radiation Biology," p. 241. Williams & Wilkins, Baltimore, Maryland.

Bacq, Z. M., and Alexander, P. (1961). "Fundamentals of Radiobiology," 2nd ed. Pergamon, Oxford.

Berger, J. D., and Kimball, R. F. (1964). *J. Protozool.* **11**, 534.

Boyle, J. M., Paterson, M. C., and Setlow, R. B. (1970). *Nature (London)* **226**, 708.

Bruce, A. K., and Malchman, W. H. (1965). *Radiat. Res.* **24**, 374.

Brunk, C. F. (1970). *Biophys. Soc. Abstr.* p. 175a.

Brunk, C. F., and Hanawalt, P. C. (1967). *Science* **158**, 663.

Brunk, C. F., and Hanawalt, P. C. (1969). *Radiat. Res.* **38**, 285.

Buetow, D. E. (1970). Personal communication.

Burchill, B. R., and Rustad, R. C. (1969). *J. Protozool.* **16**, 303.

Carrier, W. L., and Setlow, R. B. (1971). *In* "Methods in Enzymology. Nucleic Acids Part D" (L. Grossman and K. Moldave, eds.), p. 230. Academic Press, New York.

Christensen, E., and Giese, A. C. (1956). *J. Gen. Physiol.* **39**, 513.

Cook, J. S. (1967). *Photochem. Photobiol.* **6**, 97.

Cook, J. S. (1970). *In* "Photophysiology" (A. C. Giese, ed.), Vol. 5, p. 191. Academic Press, New York.

Dan, K. (1966). *In* "Cell Synchrony" (I. L. Cameron, and G. M. Padilla, eds.), p. 307. Academic Press, New York.

Dean, C. J., Feldschreiber, P., and Lett, J. T. (1966). *Nature (London)* **209**, 49.

De Lucia, P., and Carins, J. (1969). *Nature (London)* **224**, 1164.

Editorial (1969). *Nature (London)* **224**, 412.

Eichel, H. J., and Roth, J. S. (1953). *Biol. Bull.* **104**, 351.

Elliott, A. M., Brownwell, L. E., and Gross, J A. (1954). *J. Protozool.* **1**, 193.

Francis, A. A., and Whitson, G. L. (1969). *Biochim. Biophys. Acta* **179**, 253.

Frankel, J. (1962). *C. R. Trav. Lab. Carlsberg* **33**, 1.

Gibor, A. (1969). *J. Protozool.* **16**, 190.

Giese, A. C. (1945a). *Physiol. Zool.* **18**, 223.

Giese, A. C. (1945b). *J. Cell. Comp. Physiol.* **26,** 47.
Giese, A. C. (1953). *Physiol. Zool.* **26,** 1.
Giese, A. C. (1967). *Res. Protozool.* **2,** 267.
Giese, A. C. (1968). *Int. Congr. Photobiol., 5th, 1968* p. 121.
Giese, A. C., and Lusignon, M. (1960). *Science* **132,** 806.
Giese, A. C., McCaw, B. K., Parker, J. W., and Recher, J. T. (1965). *J. Protozool.* **12,** 171.
Goldfeder, A. (1963). *Laval Med.* **34,** 12.
Goutier-Pirotte, M., and Thonnard, A. (1956). *Biochim. Biophys. Acta* **22,** 396.
Grady, L. J., and Pollard, E. C. (1967). *Biochim. Biophys. Acta* **145,** 837.
Hagen, V., Ernst, H., and Cepicka, I. (1963). *Biochim. Biophys. Acta* **74,** 598.
Hanawalt, P. C., Cooper, P., and Kanner, L. (1970). *Biophys. Soc. Abstr.* p. 23a.
Hunter, N. W. (1965). *Physiol. Zool.* **38,** 343.
Iverson, R. M., and Giese, A. C. (1957). *Exp. Cell Res.* **13,** 213.
Jagger, J. (1967). "Introduction to Research in Ultraviolet Photobiology." Prentice-Hall, Englewood Cliffs, New Jersey.
Kanazir, D. T. (1969). *Progr. Nucl. Acid Res. Mol. Biol.* **9,** 117.
Kelly, R. B., Atkinson, M. R., Huberman, J. A., and Kornberg, A. (1969). *Nature (London)* **224,** 495.
Kelner, A. (1949). *Proc. Nat. Acad. Sci. U. S.* **35,** 73.
Kimball, R. F. (1955). *Radiat. Biol.* **3,** 285.
Kimball, R. F. (1964). *Biochem. Physiol. Protozoa* **3,** 243.
Kimball, R. F. (1966a). *Advan. Radiat. Biol.* **2,** 135.
Kimball, R. F. (1966b). *Radiat. Res., Suppl.* **6,** 51.
Kimball, R. F., and Gaither, N. (1951). *J. Cell. Comp. Physiol.* **37,** 211.
Kimball, R. F., Geckler, R. P., and Gaither, N. (1952). *J. Cell. Comp. Physiol.* **40,** 427.
Kollman, G., Martin, D., and Shapiro, B. (1969). *Int. J. Radiat. Biol.* **16,** 121.
Lyman, H., Epstein, H. T., and Schiff, J. A. (1961). *Biochim. Biophys. Acta* **50,** 301.
McGrath, R. A., Williams, R. W., and Swartzendruber, D. C. (1966). *Biophys. J.* **6,** 113.
Marmur, J. (1961). *J. Mol. Biol.* **3,** 208.
Marmur, J., Anderson, W. F., Matthews, L., Berns, K., Gajewska, E., Lane, D., and Doty, P. (1961). *J. Cell. Comp. Physiol.* **58,** Suppl. 1, 33.
Matsuoka, S. (1967a). *Bull. Inst. Chem. Res., Kyoto Univ.* **45,** 21.
Matsuoka, S. (1967b). *Bull. Inst. Chem. Res., Kyoto Univ.* **45,** 30.
Meselson, M., and Stahl, F. W. (1958). *Proc. Nat. Acad. Sci. U. S.* **44,** 671.
Moseley, B. E. B., and Laser, H. (1965). *Proc. Roy. Soc. Ser., B* **162,** 210.
Mossman, K. L., and Whitson, G. L. (1971). Radiation effects (in press).
Okada, S. (1969). *In* "Radiation Biochemistry" (K. I. Altman, G. B. Gerber, and S. Okada, eds.), Vol. 1, p. 148. Academic Press, New York.
Pearson, M., and Johns, H. E. (1966). *J. Mol. Biol.* **20,** 215.
Pettijohn, D., and Hanawalt, P. (1964). *J. Mol. Biol.* **9,** 395.
Pollard, E. C., and Achey, P. M. (1964). *Science* **146,** 71.
Revesz, L., and Modig, H. (1966). *Ann. Med. Exp. Biol. Fenn.* **44,** 333.
Revesz, L., Bergstrand, H., and Modig, H. (1963). *Nature (London)* **198,** 1275.
Rinaldi, A. R. (1959). *Exp. Cell Res.* **18,** 70.
Roth, J. S. (1962). *J. Protozool.* **9,** 142.
Roth, J. S., and Buccino, G. (1965). *J. Protozool.* **12,** 432.

Roth, J. S., and Eichel, H. J. (1955). *Biol. Bull.* **108**, 308.
Royal, K. M., Whitson, G. L., and Carrier, W. L. (1969). Annual Progress Report ORNL-4412, p. 63.
Rupert, C. S. (1960). *J. Gen. Physiol.* **43**, 573.
Rupert, C. S. (1962). *J. Gen. Physiol.* **45**, 703.
Scher, S., and Sagan, L. (1962). *J. Protozool.* **8**, Suppl., 8.
Schiff, J. A., and Epstein, H. T. (1968). *In* "The Biology of *Euglena*" (D. E. Buetow, ed.), Vol. 2, p. 285. Academic Press, New York.
Setlow, J. K., and Setlow, R. B. (1963). *Nature* (*London*) **197**, 560.
Setlow, R. B., and Carrier, W. L. (1964). *Proc. Nat. Acad. Sci. U. S.* **51**, 226.
Setlow, R. B., and Carrier, W. L. (1966). *J. Mol. Biol.* **17**, 237.
Setlow, R. B., Carrier, W. L., and Bollum, F. J. (1964). *Biochim. Biophys. Acta* **91**, 446.
Setlow, R. B., Carrier, W. L., and Setlow, J. K. (1969). *Biophys. J.* **9**, A-57.
Shapiro, B., Kollman, G., and Asnen, J. (1966). *Radiat. Res.* **27**, 609.
Shepard, D. C. (1965). *Exp. Cell Res.* **38**, 570.
Skreb, Y., and Bevilacqua, L. (1962). *Biochim. Biophys. Acta* **55**, 250.
Skreb, Y., and Horvat, D. (1966). *Exp. Cell Res.* **43**, 639.
Skreb, Y., Eger, M., and Horvat, D. (1965). *Biochim. Biophys. Acta* **103**, 180.
Smith, K. C. (1963). *Photochem. Photobiol.* **2**, 503.
Storrer, J. (1967). *Radiat. Res.* **31**, 699.
Stuy, J. H. (1960). *J. Bacteriol.* **79**, 707.
Stuy, J. H. (1961). *Radiat. Res.* **14**, 56.
Sullivan, W. D., and Ehrman, R. A. (1966). *Brotheria, Ser. Cienc. Natur.* **35**, 93.
Sullivan, W. D., and Sparks, J. T. (1961). *Exp. Cell Res.* **23**, 436.
Sutherland, B. M., Carrier, W. L., and Setlow, R. B. (1967). *Science* **158**, 1699.
Sutherland, B. M., Carrier, W. L., and Setlow, R. B. (1968). *Biophys. J.* **8**, 490.
Sutherland, R. M., Stannard, J. N., and Weed, R. I. (1967). *Int. J. Radiat. Biol.* **12**, 551.
Suyama, Y., and Preer, J. R. (1965). *Genetics* **52**, 1051.
Wacker, A., Dellweg, H., and Weinblum, D. (1960). *Naturwissenschaften* **20**, 477.
Watanabe, Y., and Ikeda, M. (1965). *Exp. Cell Res.* **39**, 443.
Whitson, G. L. (1965). *J. Exp. Zool.* **160**, 207.
Whitson, G. L. (1969). *J. Cell Biol.* **43**, 157a.
Whitson, G. L., and Francis, A. A. (1968a). Unpublished results.
Whitson, G. L., and Francis, A. A. (1968b). *Int. Congr. Photobiol., 5th, 1968.* p. 97.
Whitson, G. L., and Royal, K. M. (1968). Unpublished results.
Whitson, G. L., Francis, A. A., and Carrier, W. L. (1967). *J. Protozool.* **14**, Suppl., 8.
Whitson, G. L., Francis, A. A., and Carrier, W. L. (1968). *Biochim. Biophys. Acta* **161**, 285.
Whitson, G. L., Fisher, W. D., and Francis, A. A. (1969). Annual Progress Report ORNL-4412, p. 63.
Wulff, D. L., and Rupert, C. S. (1962). *Biochem. Biophys. Res. Commun.* **7**, 237.
Zeuthen, E. (1964). *In* "Synchrony in Cell Division and Growth" (E. Zeuthen, ed.), p. 99. Wiley, New York.

Chapter 5

Techniques for the Analysis of Radiation-Induced Mitotic Delay in Synchronously Dividing Sea Urchin Eggs

RONALD C. RUSTAD

I. Introduction

Sea urchin eggs offer certain special advantages for the quantitative study of radiation-induced mitotic delay. First, the natural degree of mitotic synchrony is much higher than that of yeasts, protozoans, or mammalian cells that have been "partially synchronized" by mechanical selection, temperature shocks, or metabolic manipulations (cf. Zeuthen, 1964; Cameron and Padilla, 1966; Petersen *et al.*, 1968; Stubblefield, 1968). The only comparable synchrony of mitosis would appear to be within individual plasmodia of slime molds such as *Physarum* (cf. Guttes *et al.*, 1961). Second, radiation may be administered not only to the fertilized diploid cell, but also to either the sperm or the unfertilized egg (even an anucleate egg). Third, the egg does not have to double its mass during every division cycle. Since its primary activities involve reproducing chromosomes and centrioles, this cell may be regarded as a relatively simple "mitotic machine," which even seems to have the messenger RNA's for essential protein synthesis preformed before fertilization (reviewed by Gross, 1967).

The general radiation responses during the mitotic cycle of the sea urchin egg have been reviewed recently (Rustad, 1971a), and a comparison between the effects of radiation and of various chemical inhibitors on mitosis is in preparation (Rustad, 1971b). The more specialized purposes of this contribution are to direct the attention of people concerned with radiation biology to the kinds of problems that can be submitted to quantitative analysis with synchronously dividing sea urchin eggs and to illustrate some of the special laboratory techniques that have been developed for exploiting this experimental material. Whenever possible, references concerning any particular method will be made to original papers or to a more extensive review. However, a few unpublished innovations have arisen in our laboratory, and some of the methods were simply passed on as a "folk tradition" by fellow echinoderm workers, particularly by Professors Daniel Mazia and Charles B. Metz.

II. General Procedures for Handling Gametes and Zygotes

The usual techniques for the experimental use of sea urchin eggs will be outlined briefly in order to familiarize the nonspecialist with the system. More detailed descriptions of the general procedures are available elsewhere (e.g., Harvey, 1956; Costello *et al.*, 1957; Hinegardner, 1967, 1971).

A. Obtaining and Maintaining Animals

Ideal conditions for studying sea urchin eggs are found at marine laboratories equipped with running seawater. However, animals can be shipped by air, and it is not unduly expensive or difficult to work with gametes anywhere with adequate airline connections. At present, daily quantitative experiments cannot always be planned, because breeding cycles are seasonal, and the animals may be injured in shipping and/or become unhealthy while kept in aquaria.

The duration of the experiments and the species of the animals used determine the complexity of the problem of maintaining the animals. Large-scale biochemical experiments can be performed immediately on freshly shipped animals. Usable gametes can be obtained from *Strongylocentrotus purpuratus* that have been left for several days in a moist gunnysack in a cold room. The first prerequisite for keeping the animals in aquaria over extended periods of time is a good synthetic

TABLE I

List of Suppliers of Marine Animals in North America[a,b]

Maritime Biological Laboratories Box 749 St. Stephen, New Brunswick, Canada	Tropical Atlantic Marine Specimens P.O. Box 62 Big Pine Key, Florida 33043
Gulf of Maine Biological P.O. Box 538 Brunswick, Maine 04011	Gulf Specimen Company, Inc. P.O. Box 237 Panacea, Florida 32346
Marine Biological Laboratory Woods Hole, Massachusetts 02543	G.W. Noble 3601 Biltmore Drive Panama City, Florida 32401
The Northwest Marine Specimen Co. Woods Hole, Massachusetts 02543	Pacific Bio-Marine Supply Co. P.O. Box 536 Venice, California 90291
New England Biological Associates, Inc. Sand Hill Cove Road Narragansett, Rhode Island 02882	Peninsula Marine Biologicals 1005 Benito Avenue Pacific Grove, California 93950
The Harborton Marine Laboratory Box 11 Harborton, Virginia 23389	Marino's Aquatic Specimens 301 Evergreen Avenue Daly City, California 94015
Marathon Specimens Unlimited, Inc. Box 877 Marathon, Florida 33050	

[a] Compiled by R. K. Josephson in 1969.

[b] The individual suppliers should be contacted concerning what times of the year they might provide sea urchins or sand dollars containing gametes "suitable for research" and not simply "adequate for classroom demonstrations."

seawater, such as Instant Ocean (Aquarium Systems, 1450 East 289 Street, Wickliffe, Ohio, 44092, who also provide a list of current suppliers of marine animals). In 1969, a comprehensive list of suppliers was compiled by my colleague Professor Robert K. Josephson (Table I). The choice of an aquarium depends on the nature of the experiments. Some species can be held in a cooled aquarium without filtration for a week or more, provided air is bubbled through the seawater. Continuous maintenance of small colonies of 50 to 100 animals is theoretically possible in refrigerated recirculating aquaria (e.g., Aquarium Systems), if the animals are fed agar egg cakes or suitable marine plants (see Hinegardner, 1967, 1971).

B. Obtaining Gametes

Sea urchins have five gonopores, one associated with each ovary or testis (Fig. 1). The animals can be induced to spawn by injecting 0.55 *M* KCl into the body cavity through the soft tissue between the five teeth and the edge of the test (or "shell") (Fig. 1). A small injection (e.g., 0.5 ml) can sometimes be used to identify the sex of the animals by the appearance of either white sperm or pigmented eggs at the gonopores. In some genera, notably *Arbacia,* electrical stimulation with a 6–20 V alternating current across any two gonopores or across the whole animal will induce some shedding, and a continued application of the current will lead to the discharge of most of the ripe gametes. With limited stimuli it is possible to use some of the several million eggs from the same female in a series of successive experiments.

A simple method for obtaining most or all of the mature gametes is to remove the Aristotle's lantern (the complex structure associated with the teeth) and the intestine, drain the body fluid, and then nearly fill the body cavity with 0.55 *M* KCl. Caution must be exercised to avoid spilling KCl on the outside of the animal as it can cause the release of a dermal secretion which inhibits fertilization.

The concentrated sperm are shed "dry" into a dish or removed from the aboral surface with a Pasteur pipette. Any clear fluid should be pipetted off and discarded. Sperm will survive for a week or more in a refrigerator, if they are kept in a covered dish (sealed with Parafilm to prevent drying).

The delicate eggs should be shed into seawater without passing through an air–water interface. The animal may be placed on top of a nearly full beaker of seawater. If stimulated strongly or repeatedly, the animals may shed for a half hour or more. Although some urchins may survive several

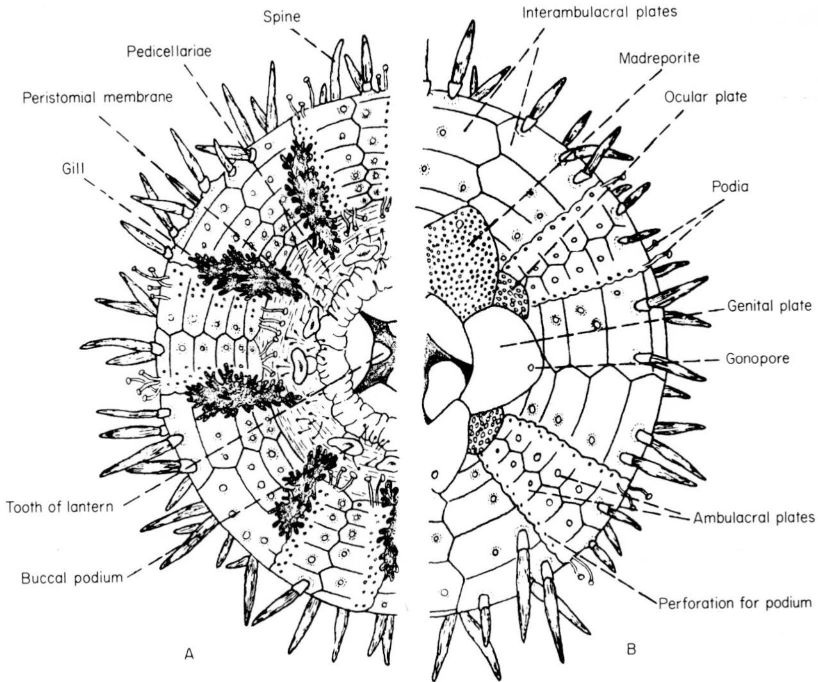

Fig. 1. The external anatomy of the sea urchin *Arbacia punctulata*. A. Oral view. B. Aboral view. The animal has five teeth and five gonopores. (After Petrunkevitch, from W. M. Reid, 1950.)

hours out of water, it is sometimes helpful to submerge the animal and the collecting beaker in another larger beaker if the female is to be used again. For optimum results, eggs are usually washed three times by stirring them in fresh seawater, allowing them to settle by gravity, and decanting off the supernatant seawater. If spines and/or feces are present, the eggs can be decanted into a fresh beaker after the heavy objects settle. If other detritus is present, it usually can be removed by filtering the eggs through bolting silk or wiping tissues. Eggs should generally be used within a few hours after shedding.

C. Fertilization and Normal Development

An approximate rule of thumb for obtaining 100% fertilization without polyspermy is to add 1 drop of undiluted "dry sperm" to 10 ml of seawater, mix thoroughly until no white aggregates are visible, add 1 drop of this sperm suspension to 10 ml of egg suspension, and mix thoroughly.

After one or two minutes the fertilized eggs will have elevated a fertilization membrane (Fig. 2). Adequate aeration may be achieved either by continuous gentle stirring of dense egg suspensions or by using fewer eggs than the number that would cover the bottom of the incubation dish.

The first synchronous cell division will occur within 45 to 120 min after fertilization depending on the species and the temperature for normal development. The subsequent mitotic cycles are considerably shorter (20 to 50 min). Eggs from some genera (e.g., *Arbacia* and *Lytechinus*) develop at room temperature, while others must be cooled (e.g., the division of *Strongylocentrotus purpuratus* is most reliable near 15°C). Figure 3 shows an example of the timing of the mitotic stages in the first division cycle of *Arbacia* eggs.

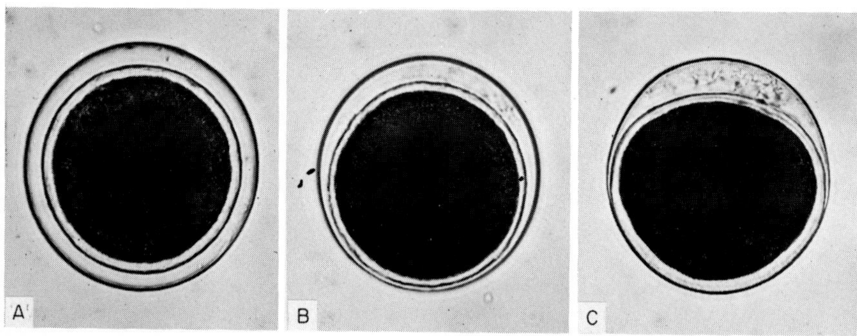

Fig. 2. Fertilized eggs of the sea urchin *Strongylocentrotus purpuratus*. A. A normal egg that has elevated its fertilization membrane (the outer dark line) and differentiated its hyaline layer (the thin clear layer near the dark central cytoplasm). B. An egg in which the elevation of the fertilization membrane has been partially inhibited by prefertilization irradiation with UV from the direction of the bottom of the page. C. An egg exhibiting complete inhibition of the elevation of the fertilization membrane on the hemisphere of the egg that had been directly exposed to a larger dose of UV-radiation. Membrane elevation occurred on the side of the egg that was shielded by the strongly UV-absorbing cytoplasm. (From Rustad, 1959a.)

III. Irradiation Techniques

The most general requirement for the quantitative study of radiation-induced mitotic delay is that each cell receive the same dose of radiation. In addition, since eggs exhibit a rapid time-dependent recovery from the effects of ionizing radiation, it is desirable to deliver the radiation rapidly to all cells at the same time. While there is no dark recovery from UV-

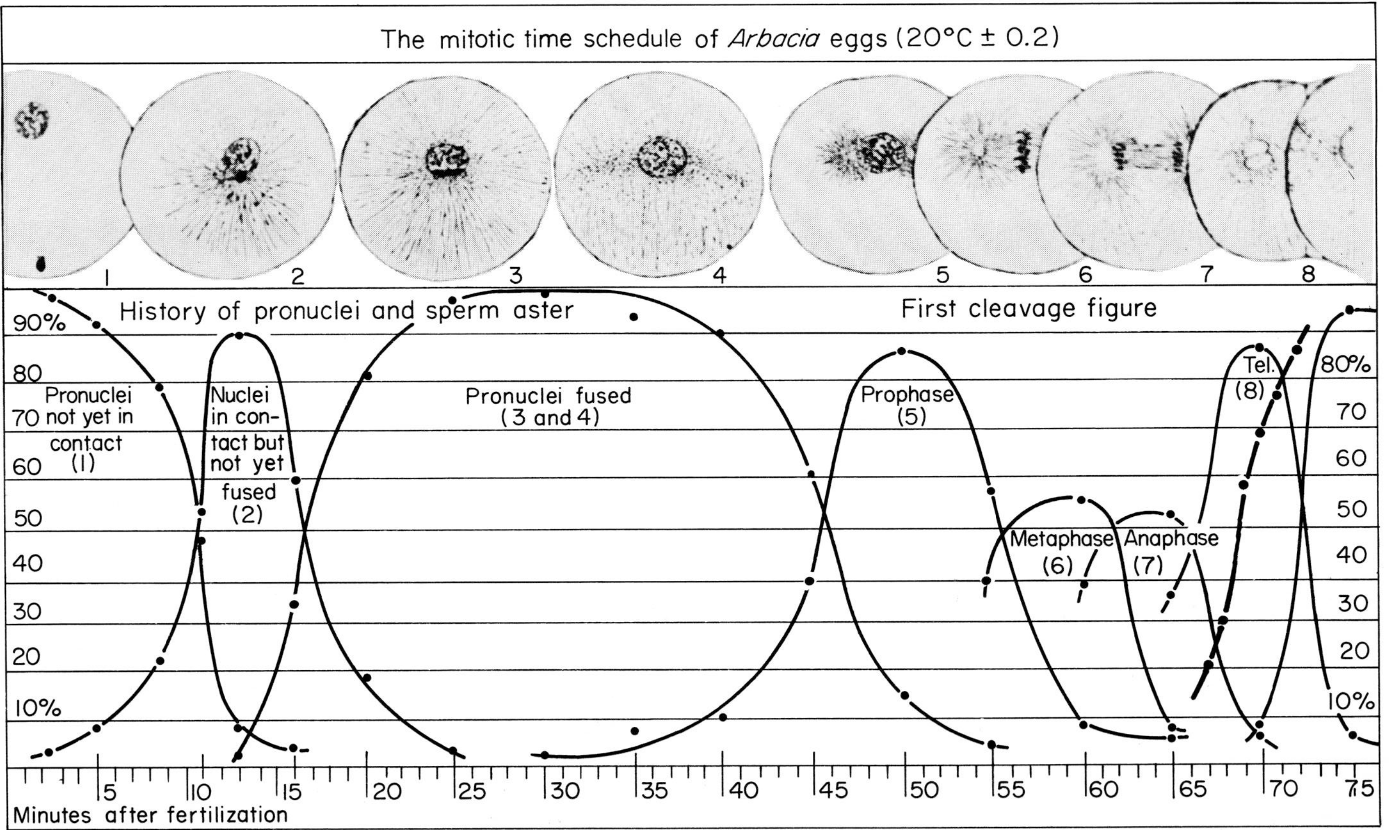

Fig. 3. Fry's mitotic time schedule. The percentage of *Arbacia* eggs in each mitotic stage is plotted against the time after fertilization. The percentage of eggs that have divided is indicated by the heavier line. At higher temperatures the first mitotic cycle may be as short as 45 min. (From Fry, 1936.)

radiation damage, eggs are photoreactivable; hence, UV experiments should be performed in the dark or with nonphotoreactivating red or yellow lighting.

A. Ionizing Radiation

Sea urchin gametes are relatively resistant to γ-radiation. This resistance appears to be due to the repair systems which are more efficient than those found in sensitive cells (such as mammalian cells in culture) (see review by Rustad, 1971a). The rapid recovery necessitates the use of X-ray or γ-ray sources with a high dose rate (see Henshaw, 1932; Rustad, 1970). For rapidly dividing genera (such as *Arbacia*) it is desirable to deliver exposures at dose rates in excess of 5 kR/min, because the effects of doses in the range of 5 to 25 kR are conventionally studied during a mitotic cycle of approximately 45 min in duration. Furthermore, in unfertilized eggs the half-time for recovery can be of the order of 10 min, and the effects of doses as large as 100 kR may be investigated. In some other more radiosensitive genera with longer division cycles (e.g., *Strongylocentrotus*) exposures as small as 1 kR can be profitably studied and lower dose rates are acceptable (see Rao, 1963).

An intense source of high-energy monochromatic γ-radiation will ensure a high dose rate and eliminate most of the problems concerning uniformly irradiating a number of dishes of cells simultaneously. The most desirable sources of ionizing radiation seem to be high-energy γ-rays from ^{60}Co or ^{137}Cs. Nonetheless, some 250 keV X-ray machines are quite satisfactory. Since the intensity is inversely proportional to the square of the distance from the X-ray target to the specimen, clinical or commercial machines may be modified to produce higher dose rates by removing equipment (such as aiming devices) that normally prevents the specimen from being moved near the tube. However, the area and uniformity of the field must then be studied with due consideration given to filtration of soft X-rays.

Whatever the choice of radiation source, the geometry of the irradiation vessels should allow uniform exposure to radiation and oxygenation of the cells. If long exposures are employed, it may be necessary to control the temperature.

The effects of both α- and β-particles on sea urchin gametes have been studied (e.g., Miwa *et al.*, 1939a,b; Failla, 1971). The dosimetry problems with these short-ranged, highly ionizing particles are so restrictive that before planning experiments the nonspecialist should consult a radiation physicist who has had prior experience with such particles.

B. UV-Radiation and Photoreactivation

The apparatus for UV-radiation experiments can be simple and inexpensive. All that is actually required is a germicidal lamp (which delivers some 88% of its UV-radiation in the 254-nm band), protection for the operators' eyes and skin, and some means of preventing photoreactivation during the experiment. An irradiator may consist of 15-in. germicidal lamps (e.g., General Electric 15W, G15T8) mounted in an ordinary fluorescent desk lamp taped to an opening cut in a cardboard box; however, an irradiator built of wood or plastic with a simple sliding shutter is preferable. Many methods are available for determining the dose rate (see Jagger, 1967), and, if an exact knowledge of exposure is not essential, the manufacturer's specifications are probably adequate for relatively new lamps used after a warm-up period of several minutes.

Clean seawater is essentially transparent to 254 nm UV-radiation and glass is opaque. Therefore, there are no geometrical exposure problems if the cells are irradiated in large open dishes at some distance from the UV lamp. However, the cells themselves absorb UV strongly, so it is absolutely essential that sperm be sufficiently diluted so that those near the bottom of the dish are not shielded, and that stationary eggs be irradiated in a monolayer. When sperm are to be irradiated, they can be diluted to the concentration used for fertilization and exposed to UV-radiation just before eggs are added to the dishes of sperm. When eggs are to be irradiated, a quantity of cells that will cover less than 50% of the bottom of the dish is stirred up and allowed to settle before the shutter of the UV-irradiator is opened. The eggs absorb so strongly that different regions of the cytoplasm receive quite different amounts of UV-radiation. A series of measurements on *Strongylocentrotus purpuratus* indicated that nearly all of the incident 254-nm UV-radiation is absorbed by an egg (Giese, 1938a). One consequence of this cytoplasmic absorption sometimes seen with large doses is a unilateral inhibition of the elevation of the fertilization membrane [which can be prevented on the side of the egg that faces the UV lamp (Spikes, 1944; Rustad, 1959a)] (Fig. 2).

Fluorescent lamps or special "black light" lamps (e.g., General Electric F40 T12/BL) provide adequate sources of photoreactivating radiation (see Jagger, 1967). A pair of ordinary 15-W tubes at a distance of 30–40 cm can produce maximum photoreactivation within 10–20 min.

To prevent photoreactivation, photobiologists often equip their laboratories with two sets of room lights: conventional fluorescent lamps and nonphotoreactivating yellow lamps (e.g., General Electric F30 T8/G0). Red darkroom safelights may also be satisfactory, and may be tested by comparing UV-radiation effects on eggs kept in total darkness against the effects on cells receiving the maximum possible red light exposure. Similarly, such a bioassay for freedom of photoreactivating wavelengths should be applied to any microscope lamp filter used for the examination of eggs that will be studied further.

IV. Quantitative Analysis of Mitotic Delay

A. Populations of Eggs

One of the reasons for the popularity of sea urchin eggs for quantitative studies of mitosis is the ease with which the division time can be determined. Division can be studied conveniently with either a dissection microscope or a compound microscope at magnifications between 25 and 200. The simplest criterion for division is any sign of furrow formation in 50% of the eggs (the furrow forms during telophase as nuclear membrane material appears around each chromosome giving the reconstituting nuclei the appearance of a grape cluster). Frequently, the furrow will begin at one side of the equatorial plate before appearing on the opposite side. For many purposes it is possible to observe a sample of several thousand eggs and estimate the percentage of cells which have divided. If the eggs divide synchronously, the observer can easily judge when there is less than 50% division and a few minutes later observe that the number divided is clearly in excess of 50%. With *Arbacia* eggs the usual accuracy of estimating the time when 50% division has occurred is about ±1 min. Two independent observers can estimate the 50% division time with the same precision (e.g., Failla and Rustad, 1970). In contrast, the subjective criteria for judging the onset of earlier mitotic stages *in vivo* lead to much greater variations in the data (see Blum *et al.*, 1954).

Division in some populations of eggs is not always synchronous. Some species show greater variability in division times than others and the physiological state of a particular animal may modify the spread of division times. UV-irradiated eggs exhibit a dose-dependent increase in the spread of division times (Wells and Giese, 1950; Rustad and Mazia, 1971). Cells irradiated during periods of the mitotic cycle when rapid variations in sensitivity to X-irradiation occur show a greater spread in division times which may reflect the fact that each individual cell is a few minutes earlier or later in its mitotic cycle than some others in the population (Rustad, 1970).

The behavior of a population may be determined by taking frequent

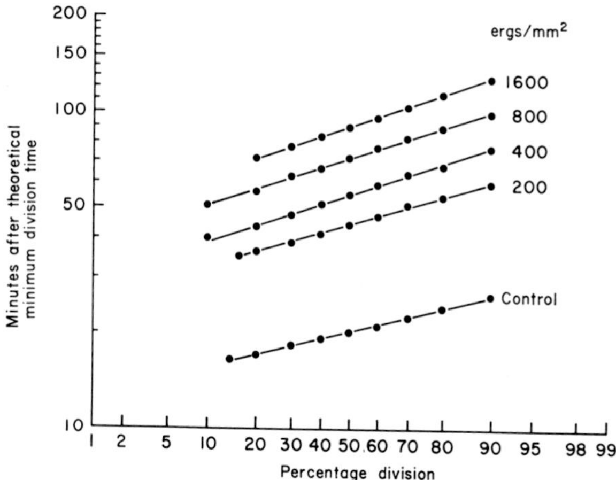

Fig. 4. The lognormal distribution of division times of *Strongylocentrotus* eggs irradiated with different doses of UV-radiation. This graphic analysis suggests that both the average division time and the loss of division synchrony are proportional to the logarithm of the dose of UV-radiation in the range between 200 and 1600 ergs/mn². (From Rustad and Mazia, 1971.)

samples and either directly counting the percentage furrowing on one or two groups of 100 cells or fixing the samples in 2–10% neutral formalin in seawater for later counting. If the division percentage is plotted against the time after fertilization, the time of 50% division can be estimated accurately. In addition, any subpopulations exhibiting different sensitivities will be detected, and the data can be submitted to conventional statistical analyses. In some instances, we have found that the distribution of division times in populations of both untreated and irradiated eggs is lognormal rather than normal (Fig. 4) (e.g., Rustad and Mazia, 1971).

B. Individual Eggs

Some workers have preferred time-lapse photography for recording division data (e.g., Blum *et al.*, 1950; Wells and Giese, 1950). Photography offers both disadvantages and advantages in comparison to the simple sampling methods. The most obvious disadvantage is that the number of cells that can be conveniently photographed is between 100 and 200, so the statistical sample of a population that can be studied is much smaller than the population that can be fixed and counted later. Earlier workers have often used "matched microscopes" to record the behavior of control and irradiated eggs: This technique requires careful attention to controlling temperature and light exposure and necessarily limits the number of experimental fields that can be studied. However, it would seem possible to construct a simple device to permit the same fields in a large number of microscope slides to be photographed repeatedly through the same microscope at regular intervals.

A major advantage of photographic recording is that the progeny of each cell may be followed through several divisions. This permits an analysis of residual radiation effects in successive mitoses (e.g., Giese, 1938; Blum *et al.*, 1950). In addition, it has been noted that the distribution of intervals between the first and second division is somewhat more uniform than the distribution of intervals between fertilization and the first division (Blum and Price, 1950). Thus, the position of an individual cell within the mitotic cycle can be more accurately determined during the second cycle than during the first, and more precise studies on cyclical variations in sensitivity may be possible if adequate numbers of cells can be photographed.

Photographic techniques are not always essential for single cell studies. If the eggs are not disturbed, the data can be recorded with the aid of a camera lucida simply by writing down the time of division of each egg on the piece of paper on which the field is projected.

V. Special Techniques

A. Prefertilization Recovery

1. *Following Irradiation*

Unfertilized eggs exhibit a time-dependent recovery from the damage of X-radiation that leads to mitotic delay (Henshaw, 1932), to direct multipolar mitoses (Rustad, 1959b), and to abnormal embryonic development (Henshaw, 1940a). Thus, it is possible to study the repair of radiation damage leading to mitotic delay in cells that are not moving through the mitotic cycle. This technique may permit the isolation of activities concerned with recovery from those concerned with ordinary progress in the mitotic cycle.

The recovery from radiation damage leading to mitotic delay is temperature-dependent (Henshaw, 1940b), but continues for some time under anaerobic conditions (Failla, 1962). The prefertilization recovery can be described by a time-dependent, one- or two-component exponential "decay" of the amount of mitotic delay (Henshaw, 1932; Miwa *et al.*, 1939a; Failla, 1962). Irradiation stimulates the incorporation of amino acids into proteins in the nearly quiescent unfertilized eggs (Rustad,

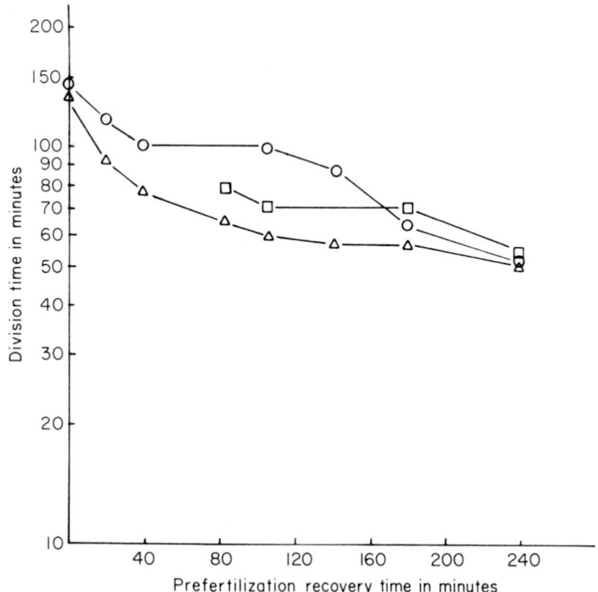

Fig. 5. The influence of puromycin on the prefertilization recovery from γ-radiation-induced mitotic delay in *Arbacia* eggs. The normal recovery is shown by triangles. Circles represent eggs which were exposed to puromycin and then washed before irradiation. Squares indicate eggs which were irradiated, then exposed for some time to puromycin, washed, and further incubated in seawater before fertilization. Puromycin seems to have little effect on the initial recovery but delays further recovery for a period approximately equal to the exposure time. Sometimes, as shown here, recovery is subsequently accelerated. (McGurn and Rustad, 1971; cf. Rustad *et al.*, 1966.)

1967). Puromycin, an inhibitor of protein synthesis, can block the slow component of the recovery (Rustad *et al.*, 1966) (Fig. 5). Thus, protein synthesis may be specifically involved in prefertilization recovery (cf. Rustad and Burchill, 1966; Failla and Rustad, 1970 for postfertilization recovery). The study of the behavior of nuclear DNA during recovery may prove difficult, because the unfertilized egg is relatively impermeable, and the DNA is very dilute with respect to the large nuclear volume and cytoplasmic mass. We have been unsuccessful in attempts to demonstrate unscheduled DNA synthesis autoradiographically. Biochemical experiments employing purified nuclei might be possible if isotopes were administered before ovulation. High levels of intracellular ^{3}H-TdR or BUdR would be needed for the demonstration of unscheduled non-semiconservative DNA synthesis (i.e., repair synthesis). If the DNA could be labeled with ^{3}H-TdR during oogenesis it would be possible to study the formation and possible excision of UV-induced pyrimidine dimers in DNA.

The procedure for studying prefertilization recovery is quite simple. The principal requirement for a study of several hours duration is that the temperature be kept constant. A dish of eggs is irradiated and aliquots are transferred to other dishes and fertilized at various times after irradiation. The eggs from some females recover quite rapidly during the first 10 to 30 min (see Failla, 1962), so samples should be fertilized at frequent intervals during the early part of the experiment. Recovery continues slowly for several hours. During the course of such experiments the division time of unirradiated eggs may decrease a few minutes.

2. *Following Artificial Stimulation*

Recently, we have found that the stimulation of unfertilized eggs with either high hydrostatic pressure or chemicals can reduce or even eliminate radiation-induced mitotic delay (e.g., Rustad *et al.*, 1969; Rustad and Greenberg, 1970; see Rustad, 1971a) (Fig. 6). The mode of action of the high hydrostatic pressure is unknown. Some evidence is consistent with the view that the effect of the detergent sodium lauryl sulfate may be to stimulate the synthesis of proteins required for the repair of the radiation-induced lesions responsible for mitotic delay.

The stimulation of eggs with a variety of chemical agents can induce activation of the fertilization reaction and even parthenogenetic development (see Kojima, 1969a). Milder treatments, such as those used in radiation studies, cause no morphological changes but can shorten the duration of the first postfertilization mitosis (e.g., Kojima, 1967, 1969b). Such treatments do not trigger DNA synthesis (Uto and Sugiyama, 1969), but they do stimulate both respiration (Sugiyama *et al.*, 1969) and the incorporation of amino acids into the proteins of unfertilized eggs (Nakashima and Sugiyama, 1969; Rustad and Greenberg, 1970) and of recently fertilized eggs (Rustad and Greenberg, 1970).

The results of our studies on radiation-induced mitotic delay in chemically stimulated eggs have been highly variable. The eggs from each female seem to respond differently, and even eggs from the same animal exhibit different responses when used over a period of an hour or two after shedding. In order to obtain optimal results, the concentration of the stimulating substance and both the duration of the stimulus and the period of incubation between stimulation and fertilization may be varied. With *Arbacia* we have often been successful when the unfertilized eggs were stimulated with (8.6×10^{-4} *M*) sodium lauryl sulfate in seawater for 90 sec, washed in seawater, and incubated for 60 min before fertiliza-

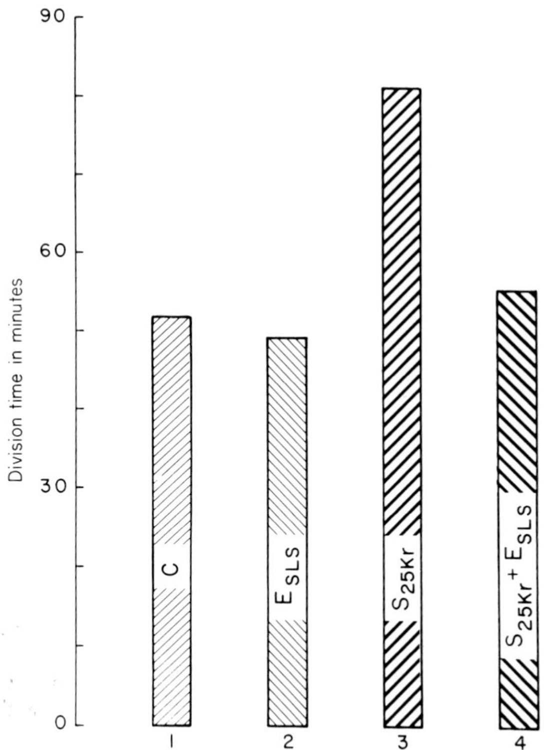

Fig. 6. The stimulation of recovery from γ-radiation-induced mitotic delay by prefertilization treatment of *Arbacia* eggs with the detergent sodium lauryl sulfate (SLS). Some of the sperm were exposed to 25 kR of γ-radiation. 1. Division time of control eggs. 2. Division time of SLS-treated eggs fertilized with normal sperm. 3. Division time of normal eggs fertilized with γ-irradiated sperm. 4. Division time of SLS-treated eggs fertilized with γ-irradiated sperm. (Rustad and Greenberg, 1969; cf. Rustad and Greenberg, 1970.)

tion. This procedure frequently leads to either partial or complete recovery from the mitotic delay that would normally arise from irradiation of the sperm, while reducing the division time of eggs fertilized with normal sperm only a few minutes. However, with some groups of eggs this treatment does not affect the mitotic delay and with other groups gross morphological changes, even lysis, occur. At the moment it is difficult to predict whether or not gentle, reproducible methods can be developed. The thresholds for successful stimulation and serious injury could be quite close as compared to the variations in sensitivity to either effect among the eggs of different animals. In point of fact, with a single stimulus the eggs from an individual female can exhibit a spectrum of

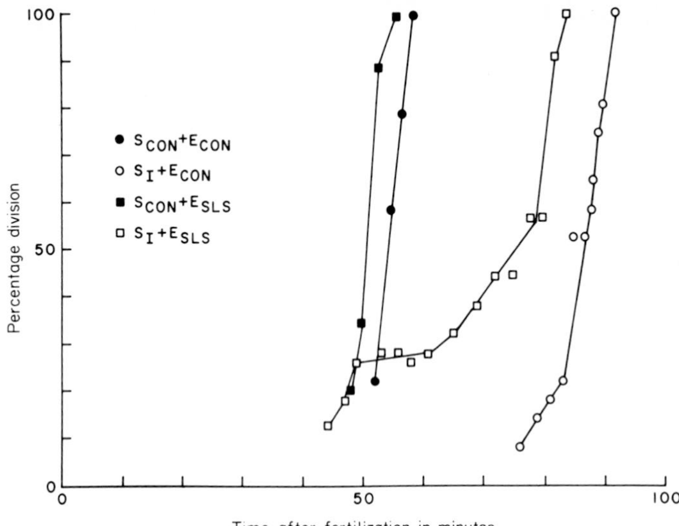

Fig. 7. Different degrees of recovery from γ-radiation-induced mitotic delay in a population of eggs from a single female *Arbacia* treated with the detergent sodium lauryl sulfate (SLS). Percentage division is plotted against time. Solid circles are the controls. Open circles indicate normal eggs fertilized with irradiated sperm. Closed squares represent SLS-treated eggs fertilized with normal sperm. Open squares show SLS-treated eggs fertilized with irradiated sperm. In this experiment it can be seen that the SLS-treatment did not enhance recovery from radiation-induced mitotic delay in approximately half of the population, induced partial recovery in one quarter of the population, and led to complete recovery in one quarter of the population. (Rustad and Greenberg, 1969; cf. Rustad and Greenberg, 1970.)

effects ranging from zero to complete recovery from radiation-induced mitotic delay (Fig. 7).

In conclusion, it should be stressed that the prefertilization stimulation techniques are highly experimental and may yield either enhanced recovery or gross damage depending on the susceptibility of the particular group of eggs studied.

B. Postfertilization Recovery

The sensitivity to X-ray-induced mitotic delay rises to a peak during the mitotic cycle (Yamashita *et al.*, 1939; Henshaw and Cohen, 1940; Rao, 1963; Failla, 1969; Rustad, 1970). This period of increasing sensitivity may reflect the time available for recovery prior to some critical mitotic stage (cf. Lea, 1946; Rustad, 1970, 1971a) (i.e., cells irradiated early in the mitotic cycle may exhibit relatively little mitotic delay because they have more time available for recovery than those irradiated later).

It should be noted that the techniques for prefertilization stimulation discussed in the preceding section might also lead to an unusually high rate of postfertilization recovery. In point of fact, preliminary measurements indicated that the prefertilization detergent treatments lead to an increased rate of amino acid incorporation into proteins not only in unfertilized eggs but also in recently fertilized ones (Rustad and Greenberg, 1970).

Failla (1962) has devised a simple technique for studying postfertilization recovery: She interrupts progress through the mitotic cycle by gently bubbling nitrogen through the medium to make the eggs anoxic. During the N_2 treatment the recovery from γ-radiation damage in the fertilized egg is more rapid than in the unfertilized egg (Failla, 1965). However, the rate of postfertilization recovery in normally oxygenated eggs is unknown. Failla's method has also proved valuable for the study of the inhibition of recovery which will be discussed in the next section.

There is no postfertilization recovery from damage to sperm by α-radiation (Failla, 1971). Postfertilization dark recovery from UV damage to sperm has been detected (Rustad and Failla, 1971).

C. Inhibition of Recovery

Low temperature strongly inhibits prefertilization recovery from radiation-induced mitotic delay (Henshaw, 1940b). This observation indicates that energy-dependent chemical reactions are essential for the recovery process. Recovery can continue for some time in anoxic eggs and in eggs incubated in the presence of respiratory inhibitors (e.g., Failla, 1962, 1969). The use of specific chemical inhibitors in attempts to deduce the nature of radiation effects is complicated by the fact that the egg contains large pools of precursors and is not especially permeable. Large cells, such as eggs and protozoans, often require drug concentrations that are several orders of magnitude higher than those used to obtain comparable effects on bacteria or cultured mammalian cells.

In eggs of sand dollars (which are closely related to sea urchins) Cook (1968) demonstrated an enhancement of UV sensitivity by 5-bromodeoxyuridine (BUdR). His results strongly suggest that some cellular DNA is the radiation-sensitive target. With eggs of the sea urchin *Arbacia punctulata* we were unable to demonstrate any reproducible enhancement of γ-ray sensitivity by BUdR (Rustad and Burchill, 1963). The reasons for this failure may emerge from continued studies on the incorporation of BUdR and TdR into DNA (e.g., Zeitz *et al.*, 1968).

Actinomycin D does not influence the magnitude of radiation-induced mitotic delay (Rustad and Burchill, 1966). This observation suggests that any essential protein synthesis is directed by long-lived messenger RNA's present in the unfertilized egg.

We have argued that protein synthesis is essential for recovery from the damage that leads to mitotic delay, because puromycin increases the magnitude of the delay when applied before fertilization (Rustad *et al.*, 1966), after fertilization (Rustad and Burchill, 1966), or during post-fertilization recovery in cells in which the mitotic cycle has been interrupted by anoxia (Failla and Rustad, 1970). The generality of our view is in harmony with recent observations on mammalian cells (Walters and Petersen, 1968; Doida and Okada, 1969; Bachetti and Sinclair, 1970) and plant meristematic cells (Van't Hof and Kovacs, 1970).

At present it is difficult to predict the future of studies on eggs with inhibitors. Large molecules may not penetrate the egg in adequate quantities during the course of short experiments. The large precursor pools may render many antimetabolites ineffective.

Fortunately, the ease with which inhibitor experiments can be performed on eggs ensures that each new inhibitor of interest can be examined without a great expenditure of time. Since dozens (or potentially hundreds) of dishes of eggs can be screened within a few hours, a variety of concentrations of several compounds of potential interest can be tested simultaneously.

Perhaps the only comment concerning techniques for inhibitor studies is to note that small dishes can be used if the available quantity of the experimental chemical is limited. Small containers should always be covered to prevent evaporation during the course of an experiment. A variety of small plastic dishes and slides are made by Falcon Plastics (5500 W. 83rd St., Los Angeles, Calif. 90045) and conventional glass depression slides are readily available from most laboratory supply houses.

D. Enucleation by Centrifugation

When sea urchin eggs are centrifuged for approximately 10 min at 10,000g, the cell organelles become stratified within the cell according to their densities (Fig. 8) (see Harvey, 1956). Higher forces and/or longer times of centrifugation cause the egg to pull apart into half- or quarter-eggs of different cytoplasmic compositions (Fig. 8). Electron micrographs have revealed that some organelles (notably mitochondria) are found

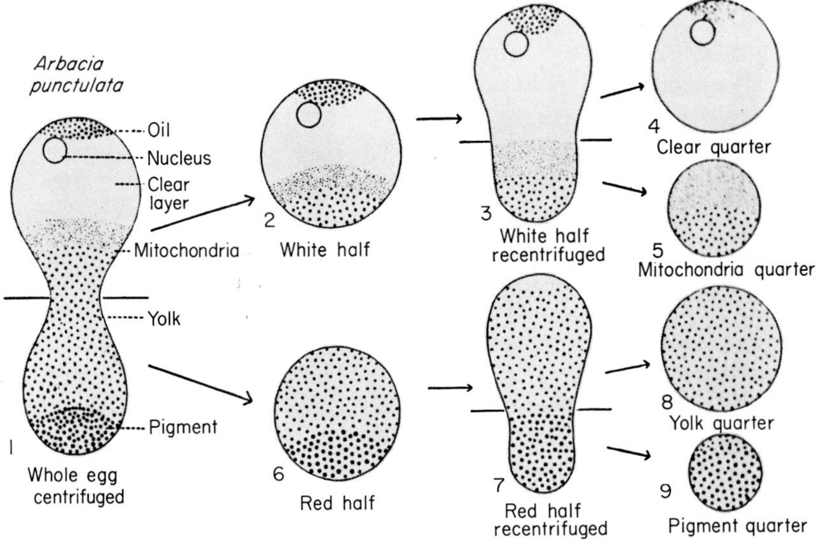

Fig. 8. Centrifugation of the unfertilized *Arbacia* egg into halves and then into quarters. At high centrifugal forces the cytoplasmic organelles become stratified according to their densities and the eggs then elongate and pinch into fertilizable, membrane-limited fragments. (From Harvey, 1956.)

throughout stratified eggs and in all fragments (Geuskens, 1965; Anderson, 1970).

The nucleate half-egg is sensitive to X-ray-induced mitotic delay while no sensitivity of the enucleate centrifuged half (subsequently fertilized with a normal sperm) has been detected (Henshaw, 1938; Blum *et al.*, 1951; Rustad *et al.*, 1963). The apparent lack of radiation sensitivity of the cytoplasm may be a consequence of the fact that the division of the haploid centrifuged half is greatly delayed compared to either unirradiated or irradiated anucleate halves that have been prepared by microsurgery and later fertilized with normal sperm (Rustad *et al.*, 1971) (see Section V,E). Nonetheless, the centrifugation techniques may be useful for *in vivo* radiation studies on cell organelles in half- or quarter-eggs or for concentrating cell organelles for selective microbeam irradiation (see Section V,F).

Arbacia eggs are conventionally placed in seawater layered over either 0.95 or 1.1 *M* sucrose [which Harvey (1956) recommends preparing in tap water rather than distilled water]. At room temperature, stratification and subsequent splitting into halves will occur within 10 to 20 min at 10,000–15,000*g*. For any species of eggs the optimal procedure may vary seasonally and even from animal to animal. The results depend on the ionic composition of the medium, rate of acceleration, time, force, and temperature (see Harvey, 1956). The results may be adversely influenced by small quantities of mold growing in a sucrose solution which has been stored for some time in a refrigerator.

If a gradient has formed at the seawater–sucrose interface, three separate cell layers can be seen: the centripetal halves (which are nucleate), the unbroken whole cells, and the pigmented, centrifugal halves (which are anucleate). With large forces, centrifuged halves or quarters will form a pellet at the bottom of the tube. Superior separations can be obtained if the cells are placed in linear gradients or if several layers of seawater–sucrose solutions are employed. Denny and Tyler (1964) recommended that for *Lytechinus pictus* the lowest layer be a 3:1 solution of 1.1 *M* sucrose and seawater (containing a dense egg suspension), the middle layer a 3:2 solution of 1.1 *M* sucrose and seawater, and the top layer seawater. For *Strongylocentrotus purpuratus* they suggested making the middle layer a 6.3:3.7 solution of 1.1 *M* sucrose and seawater. They found that the nonnucleate halves banded at the interface between the two sucrose layers while the nucleate halves (and any unbroken whole eggs) banded just below the seawater layer. Eggs of *Lytechinus variegatus* from northern Florida elongate greatly during centrifugation but do not seem to separate into halves (Metz, 1960; Rustad, 1971).

Following centrifugation the egg layers should be washed in large volumes of seawater. Prolonged exposure to sucrose can lead to activation of the fertilization membrane and may allow ions to leak out of the eggs. Some of the organelles begin to become redistributed by Brownian motion if the centrifuged eggs remain healthy, but the oil caps may persist for hours.

E. Enucleation by Microsurgery

Eggs can be cut into approximately equal halves with fine glass needles. These nucleate and anucleate halves should have similar cytoplasmic compositions in contrast to the very different cytoplasmic compositions found between cell fragments prepared by centrifugation (see Section V,D). Either the whole eggs and halves, or the sperm may be irradiated before fertilization.

In the case of γ-radiation-induced mitotic delay, comparisons between whole and nucleate and anucleate half eggs have clearly demonstrated:

1. Nucleate half-eggs are very sensitive (Rustad *et al.*, 1964, 1971) (Fig. 9).
2. Anucleate half-eggs exhibit a lesser, yet very significant sensitivity (Rustad *et al.*, 1964, 1971) (Fig. 9).

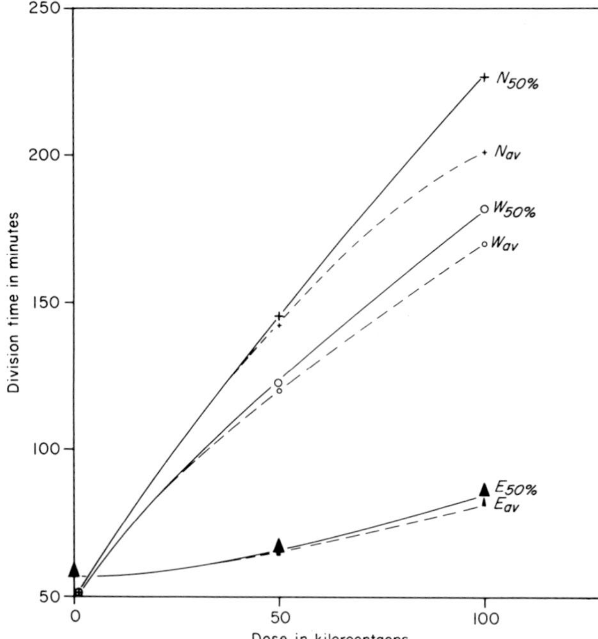

Fig. 9. The relationship of division time to the dose of γ-radiation administered to whole *Arbacia* eggs (*W*) and to nucleate (*N*) and enucleated (*E*) half-eggs following fertilization with normal sperm. The times when 50% of the eggs in each of the populations had divided are shown. In addition, the average division time (subscript *av*) of those cells in each population that were able to divide are plotted, because many cells did not divide following large doses. (From Rustad *et al.*, 1971.)

3. The expression of damage (presumably nuclear) to sperm is reduced by increased cytoplasmic volume and by increased length of the normal division cycle.

(Note: all of these observations are consistent with the hypothesis that radiation sensitivity is strongly influenced by the rate of repair and time available for recovery.) (Rustad, 1966; see also Rustad, 1971a.)

In the case of UV-induced mitotic delay, observations on whole and half-eggs have shown:

1. Nucleate half-eggs are hypersensitive (Rustad *et al.*, 1964).
2. Anucleate half-eggs can be as sensitive as whole eggs (raising the question of whether or not the damage measured in the whole egg is entirely cytoplasmic) (Rustad *et al.*, 1964).
3. The cytoplasmic damage can be photoreactivated before fertilization suggesting that a cytoplasmic DNA, perhaps mitochondrial, is the target (Rustad *et al.*, 1966).

Preparation of nucleate and anucleate half-eggs by microsurgery is not easy. Previous workers using micromanipulators only cut from one egg to as many as a dozen eggs per experiment (see Table II in Rustad *et al.*, 1970). We have found that after some practice with freehand techniques it is possible for a single individual to cut and separate from 20 to 100 eggs per hour.

The eggs are supported on an agar surface (0.2–0.3% in 80% seawater or 80% Instant Ocean), which has been previously heated and poured into the cutting dish and washed overnight in a refrigerator. Long glass microneedles can be pulled freehand from 2 mm capillary tubing over a commercial microburner or one made from a bent hypodermic needle regulated by a small clamp on the gas hose. The tip of the microneedle is rested on the agar at some distance from the egg and then the shaft of the needle is pressed slowly down on the egg, with a sawing motion if necessary (see Rustad *et al.*, 1970 for further details). The response of eggs to the needle is quite variable, and the eggs from some animals are totally unsuitable for cutting experiments. It should be cautioned that this is a highly specialized technique which an investigator may have to practice daily for several weeks or months before he achieves sufficient proficiency to prepare enough half-eggs to perform a radiation experiment.

F. Partial Cell Irradiation with Microbeams

Miwa *et al.* (1939b) deduced that the nucleus was a primary target in the sea urchin egg, because little mitotic delay arose when α-particles were unable to penetrate to the nucleus. Microbeam experiments with UV-radiation also suggest that the nucleus or some perinuclear structure is an especially radiation-sensitive target, whereas irradiating a small volume of unspecialized cytoplasm has no measurable effect (Rustad, 1964). The UV microbeam experiments also indicated that the centriole or some other component of the aster might be a radiation-sensitive component (Rustad, 1964), but definitive experiments have not been performed. Nonetheless, large doses of UV applied to a large fraction of the cytoplasm do inhibit mitosis in sand dollar eggs (Uretz and Zirkle, 1955).

Many devices have been constructed for the purpose of irradiating small regions of the cell (reviewed by Zirkle, 1957; Smith, 1964; Forer, 1966; Rustad, 1968). Any simple microbeam can be used with eggs, because the targets are relatively large. Irradiation of individual chromosomes does not appear possible because they cannot be identified routinely *in vivo* with either phase contrast microscopy or Nomarski differential interference microscopy. Since the eggs are spherical it is often desirable to flatten them in order to ensure that the beam of radiation has a short direct path to the target. Indeed, in the case of UV-radiation it is by no means certain that the visual image and the UV-projection point would be in the same place after crossing a spherical surface of unknown refractive index with respect to visible versus UV light.

Partial cell irradiation might also help to localize sensitive cytoplasmic structures in eggs in which the organelles had been stratified by centrifugation (see Section V,D). Furthermore, it should be possible to use an ordinary microscope as a visible light microbeam for the purpose of localizing photoreactiveable structures (e.g., nuclei and mitochondria) following whole-cell UV-irradiation.

G. Unusual Techniques

There are a number of unusual opportunities for analyzing particular phenomena with sea urchin eggs. Without actually outlining each of the relatively simple methods, I would like to briefly point out their existence.

1. *Induction of Cytasters*

A variety of chemical treatments lead to the production of numerous cytasters (see Harvey, 1956) which contain centrioles (Dirksen, 1961). These techniques could be useful for studying the assembly of centrioles and astral rays (microtubules).

2. *Reciliation*

Cilia are formed at the blastula stage. The cilia can be removed by treatment with hypertonic seawater, and new cilia will grow out without concomitant synthesis of new proteins (Auclair and Siegel, 1966). Reciliation occurs if the embryos are γ-irradiated either just before or just after deciliation (Rustad, 1971), but neither the effects of UV-irradiation nor the effect of irradiating quite some time before deciliation have been investigated.

3. *Cross-Fertilization*

The sperm from one genus of sea urchins usually will fertilize eggs of another genus (in fact, sea urchin–sand dollar hybrids can be formed). When suitable radiation-sensitive and radiation-resistant species are found, cross-fertilization may prove a valuable tool for analyzing a variety of problems.

4. *Polyspermy and Refertilization*

If a sperm suspension is too concentrated, the eggs are likely to become polyspermic. A variety of chemical treatments enhance the sensitivity to polyspermy. Similarly, fertilized eggs can be treated with chemicals that will permit refertilization (see Harvey, 1956). These techniques offer the possibility of making different permutations and combinations of irradiated and unirradiated eggs and/or male pronuclei. It should be possible to answer such simple questions as: "Is the ability to divide dependent on the most heavily damaged genome?" or "Do different genomes compete for the same repair enzymes?"

5. *Inhibition of Pronuclear Fusion and Cell Division*

A variety of physical and chemical treatments inhibit pronuclear fusion and/or karyokinesis and/or cytokinesis (see Zimmerman and Zimmerman, 1967, for references). Rao (1963) blocked pronuclear fusion in previously X-irradiated eggs with mercaptoethanol and obtained evidence suggestive of a transfer of cytoplasmic radiation damage to the male pronucleus. We have attempted to study recovery from radiation damage in eggs blocked at metaphase with Colcemid, but could not separate the variable mitotic delay induced by the drug from the radiation effect. We have hoped that other division-arresting agents may prove more suitable.

VI. Special Advantages and Disadvantages in Comparison with Partially Synchronized Growing Cells

The application of existing techniques does not permit the achievement of a high degree of mitotic synchrony in most growing cells (see Petersen *et al.*, 1968; Stubblefield, 1968; Bachetti and Sinclair, 1970). Furthermore, artificial synchronization of cells by chemical inhibitors may not resolve a variety of problems because cells blocked in metaphase or in nuclear S are known to continue many other activities associated with further progress in the mitotic cycle. Mechanical selection may prove the most useful synchrony method for growing cells. For lethality studies on mammalian cells that have been partly synchronized with a temporary Colcemid block, it is possible to study a more closely synchronized fraction of the population by killing the cells which incorporate tritiated thymidine when other cells have completed DNA synthesis (Sinclair and Morton, 1966). Unfortunately, this "suicide technique" cannot be used for studies on mitotic delay because the deaths may occur in later generations. In general, naturally synchronous systems such as marine eggs and syncytial slime molds seem best suited for a variety of quantitative radiation studies. As a hypothetical example: Partially synchronizing a mammalian cell culture with a generation time of 18 hr so that 80% of the cells divide over a 2-hr interval could yield some useful information about changes in radiation sensitivity over a 10-hr S period but would be completely useless for analyzing variations in sensitivity during a 30-min prophase. Substantially shorter periods of time have been studied both in sea urchin eggs (e.g., Henshaw and Cohen, 1940; Rustad, 1960, 1970; Failla, 1969) and in the slime mold *Physarum* (Devi *et al.*, 1968). As discussed above (Section IV), quantitative measurements with sea urchin eggs are easy, and even visual estimates of the division time of a population are usually accurate to within ±1 min. Hence, sea urchin eggs should remain an especially favorable material for many types of quantitative studies concerning radiation-induced mitotic delay.

In the introduction, the suggestion was made that the egg is a relatively simple mitotic machine, which does not have to reproduce itself during each division cycle. While the synthesis of DNA and some proteins are required for mitosis, the essential RNA's appear to be stored in an inactive form in the unfertilized egg (reviewed by Gross, 1967). Radiation does not appear to affect either the time of onset or the rate of nuclear DNA synthesis (Rao and Hinegardner, 1965; Zeitz *et al.*, 1968). Therefore, the radiation effects leading to the mitotic delay of eggs may be those (uniquely) involving prophase (see discussion in Rustad, 1971a). Thus, the egg is a simple cell, and we may be studying selected radiation-sensitive events. This may simplify the problem of analysis, but it may mean that other radiation-sensitive events will have to be examined in other systems.

Marine eggs also offer the advantage that the radiations can be applied separately to the haploid gametes or to the diploid zygote. In the case of sperm there is virtually no cytoplasm (even though the mitochondrion or the centriole might be a special radiation-sensitive target). It is also possible to irradiate nucleate and anucleate half-eggs prepared either by centrifugation or by microsurgery (see Sections V,D and V,E).

Marine eggs are convenient experimental materials. No elaborate equipment is needed to achieve synchronous division. The mitotic cycles of sea urchin eggs are quite short, ranging from approximately 45 min to 2 hr, and many species can be studied at room temperature. The investigator can start an experiment at the time he wishes simply by mixing suspensions of eggs and sperm. It is sometimes possible to study the gametes of the same animals again several hours after an initial experiment provided that the cells are kept refrigerated. Some species of animal, notably *Arbacia punctulata,* can be stimulated to shed gametes on a series of successive days. Some caution might be advised, however, concerning the repeated spawning, because there could be a gradient in maturity of the eggs or sperm.

One of the unique advantages of the sea urchin egg is that recovery from radiation damage can be studied outside of the mitotic cycle, because the unfertilized eggs repair damage (Section V,A). Furthermore, it is possible to artificially stimulate recovery in the unfertilized egg, so it may be possible to establish correlations between the recovery and plausible physiological control mechanisms.

The sea urchin egg is hardly a panacea for the problems of radiation biology. Many of the properties that make it useful for physiological studies are related to characteristics that make biochemical analyses difficult. First of all, the egg is a very large cell with a single normal nucleus. Hence, there is very little nuclear DNA in comparison to the volume of cytoplasm (cytoplasmic DNA, mostly mitochondrial, may account for some 80 to 95% of the cellular DNA; see Piko *et al.,* 1967). Secondly, there may be no significant RNA synthesis during the early mitoses.

The cells are not very permeable to a variety of precursors and antimetabolites, and, in general, the unfertilized eggs are much less permeable than fertilized ones. The cells also appear to have relatively large pools of a variety of amino acids and nucleic acid precursors. The combination of low permeability and large pools makes some inhibitor experiments difficult or impossible.

The high resistance of eggs to the effects of ionizing radiation necessitates the use of high dose rate sources for many types of experiments (see Section III,A). Such sources of X- or γ-radiation are expensive, so it is fortunate that UV-radiation experiments can be performed with a desk lamp and a cardboard box (Section III,B).

A final problem in working with marine eggs is the seasonal variation. Several different genera must be studied if a year-round experimental program is contemplated. More importantly, a single tropical storm or even an unusually warm or cold season can leave the investigator with no source of animals with mature gametes for a period of several weeks or even months. Continuous laboratory culture of small groups of *Lytechinus pictus* has proved possible, but the amount of gametes obtained is too small to meet all experimental requirements. At present, it would seem prudent for anyone considering a research program with sea urchin eggs to also maintain programs with other experimental materials.

VII. Conclusion

The sea urchin egg offers some unusual opportunities for quantitative studies in radiation biology, because the division is highly synchronous, the gametes may be irradiated separately, recovery occurs in unfertilized eggs outside the mitotic cycle, and the mitotic cycle may be exceptionally simple in that no mass doubling occurs and RNA synthesis does not seem essential. However, such factors as the large size of the egg, its low permeability, and its large precursor pools hamper some types of physiological and biochemical studies. Nonetheless, until the methods of artificially synchronizing exponentially growing cell populations are vastly improved, the sea urchin egg will remain one of the most useful cells for the quantitative analysis of radiation-induced mitotic delay.

Acknowledgments

Recent studies in the author's laboratory have been supported by Contract W-31-109-ENG-78 (Report No. C00-78-240) with the U. S. Atomic Energy Commission. Much of this work has been performed at the Marine Biological Laboratory, Woods Hole, Mass.

Thanks are due many people who offered constructive suggestions regarding the preparation of sections of this manuscript. I am particularly grateful to Miss A. M. Belanger, Dr. J. D. Caston, Dr. R. P. Davis, Miss J. M. Molin, Dr. O. F. Nygaard, and Dr. N. L. Oleinick.

References

Anderson, E. (1970). *J. Cell Biol.* **47,** 711–733.

Auclair, W., and Siegel, B. W. (1966). *Science* **154,** 913–915.

Bachetti, S., and Sinclair, W. K. (1970). *Radiat. Res.* **44,** 788–806.

Blum, H. F., and Price, J. P. (1950). *J. Gen. Physiol.* **33,** 305–310.

Blum, H. F., Loos, G. M., and Robinson, J. C. (1950). *J. Gen. Physiol.* **34,** 167–181.

Blum, H. F., Robinson, J. C., and Loos, G. M. (1951). *J. Gen. Physiol.* **35,** 323–324.

Blum, H. F., Kauzmann, E. F., and Chapman, G. B. (1954). *J. Gen. Physiol.* **37,** 325–334.

Cameron, I. L., and Padilla, G. M., eds. (1966). "Cell Synchrony." Academic Press, New York.

Cook, J. S. (1968). *Exp. Cell Res.* **50,** 627–638.

Costello, D. P., Davidson, M. E., Eggers, A., Fox, M. H., and Henley, C. (1957). "Methods for Obtaining and Handling Marine Eggs and Embryos." Marine Biol. Lab., Woods Hole, Massachusetts.

Denny, P. C., and Tyler, A. (1964). *Biochem. Biophys. Res. Commun.* **14,** 245–249.

Devi, J. R., Guttes, E., and Guttes, S. (1968). *Exp. Cell Res.* **50,** 589–598.

Dirksen, E. R. (1961). *J. Biophys. Biochem. Cytol.* **11,** 244–247.

Doida, Y., and Okada, S. (1969). *Radiat. Res.* **38**, 513–529.
Failla, P. M. (1962). *Science* **138**, 1341–1342.
Failla, P. M. (1965). *Radiat. Res.* **25**, 331–340.
Failla, P. M. (1969). *Radiology* **93**, 643–648.
Failla, P. M. (1971). Manuscript in preparation.
Failla, P. M., and Rustad, R. C. (1970). *Int. J. Radiat. Biol.* **17**, 385–388.
Forer, A. (1966). *Exp. Cell Res.* **43**, 688–691.
Fry, H. J. (1936). *Biol. Bull.* **70**, 89–99.
Geuskens, M. (1965). *Exp. Cell Res.* **39**, 413–417.
Giese, A. C. (1938). *Biol. Bull.* **75**, 238–247.
Giese, A. C. (1938a). *Biol. Bull.* **74**, 330–341.
Gross, P. R. (1967). *Curr. Top. Develop. Biol.* **2**, 1–43.
Guttes, E., Guttes, S., and Rusch, H. P. (1961). *Develop. Biol.* **3**, 588–614.
Harvey, E. B. (1956). "The American *Arbacia* and Other Sea Urchins." Princeton Univ. Press, Princeton, New Jersey.
Henshaw, P. S. (1932). *Amer. J. Roentgenol. Radium Ther.* [N. S.] **27**, 890–898.
Henshaw, P. S. (1938). *Amer. J. Cancer* **33**, 258–264.
Henshaw, P. S. (1940a). *Amer. J. Roentgenol. Radium Ther.* [N. S.] **43**, 913–916.
Henshaw, P. S. (1940b). *Amer. J. Roentgenol. Radium Ther.* [N. S.] **43**, 921–922.
Henshaw, P. S., and Cohen, I. (1940). *Amer. J. Roentgenol. Radium Ther.* [N. S.] **43**, 917–920.
Hinegardner, R. T. (1967). *In* "Methods in Developmental Biology" (F. H. Wilt and N. K. Wessels, eds.), pp. 139–155. Crowell-Collier, New York.
Hinegardner, R. T. (1971). *In* "The Sea Urchin Embryo" (G. Czihak, ed.). Springer-Verlag, Berlin and New York (in preparation).
Jagger, J. (1967). "Introduction to Research in Ultraviolet Photobiology." Prentice-Hall, Englewood Cliffs, New Jersey.
Kojima, M. K. (1967). *Embryologia* **10**, 75–82.
Kojima, M. K. (1969a). *Embryologia* **10**, 334–342.
Kojima, M. K. (1969b). *Embryologia* **10**, 323–333.
Lea, D. E. (1946). "Actions of Radiation on Living Cells." Cambridge Univ. Press, London and New York.
McGurn, E., and Rustad, R. C. (1971). Unpublished data.
Metz, C. B. (1960). Personal communication.
Miwa, M., Yamashita, H., and Mori, K. (1939a). *Gann* **33**, 1–12.
Miwa, M., Yamashita, H., and Mori, K. (1939b). *Gann* **33**, 323–331.
Nakashima, S. K., and Sugiyama, M. (1969). *Develop. Growth & Differentiation* **11**, 115–122.
Petersen, D. F., Anderson, E. C., and Tobey, R. A. (1968). *Methods Cell Physiol.* **3**, 347–370.
Piko, L., Tyler, A., and Vinograd, J. (1967). *Biol. Bull.* **132**, 68–90.
Rao, B. (1963). Ph.D. Thesis, University of California, Berkeley.
Rao, B., and Hinegardner, R. T. (1965). *Radiat. Res.* **26**, 534–537.
Reid, W. M. (1950). *In* "Selected Invertebrate Types" (F. M. Brown, ed.), John Wiley, New York.
Rustad, R. C. (1959a). *Biol. Bull.* **116**, 294–303.
Rustad, R. C. (1959b). *Biol. Bull.* **117**, 437.
Rustad, R. C. (1960). *Exp. Cell Res.* **21**, 596–602.
Rustad, R. C. (1964). *Photochem. Photobiol.* **3**, 529–538.
Rustad, R. C. (1966). *Biol. Bull.* **131**, 404.

Rustad, R. C. (1967). *J. Cell Physiol.* **70,** 75–78.
Rustad, R. C. (1968). *Experientia* **24,** 974–975.
Rustad, R. C. (1970). *Radiat. Res.* **42,** 498–512.
Rustad, R. C. (1971a). *In* "Developmental Aspects of the Cell Cycle" (I. L. Cameron, G. M. Padilla, and A. M. Zimmerman, eds.), pp. 127–159. Academic Press, New York.
Rustad, R. C. (1971). Unpublished data.
Rustad, R. C. (1971b). *In* "The Sea Urchin Embryo" (G. Czihak, ed.). Springer-Verlag, Berlin and New York (in preparation).
Rustad, R. C., and Burchill, B. R. (1963). Unpublished experiments.
Rustad, R. C., and Burchill, B. R. (1966). *Radiat. Res.* **29,** 203–210.
Rustad, R. C., and Failla, P. M. (1971). *Biol. Bull.* **141,** 400.
Rustad, R. C., and Greenberg, M. (1969). Unpublished data.
Rustad, R. C., and Greenberg, M. (1970). *Radiat. Res.* **43,** 270.
Rustad, R. C., and Mazia, D. (1971). Manuscript in preparation.
Rustad, R. C., Corabi, M., and Yuyama, S. (1963). *Radiat. Res.* **19,** 206–207.
Rustad, R. C., Yuyama, S., and Rustad, L. C. (1964). *Biol. Bull.* **127,** 388.
Rustad, R. C., McGurn, E., Yuyama, S., and Rustad, L. C. (1966). *Radiat. Res.* **27,** 543.
Rustad, R. C., Zimmerman, A. M., and Greenberg, M. A. (1969). *Biol. Bull.* **137,** 412–413.
Rustad, R. C., Yuyama, S., and Rustad, L. C. (1970). *Biol. Bull.* **138,** 184–193.
Rustad, R. C., Yuyama, S., and Rustad, L. C. (1971). *J. Cell. Biol.* **49,** 906–912.
Sinclair, W. K., and Morton, R. A. (1966). *Radiat. Res.* **29,** 450–474.
Smith, D. L. (1964). *Int. Rev. Cytol.* **16,** 133–153.
Spikes, J. D. (1944). *J. Exp. Zool.* **95,** 89–103.
Stubblefield, E. (1968). *Methods Cell Physiol.* **3,** 25–43.
Sugiyama, M., Ishikawa, M., and Kojima, M. K. (1969). *Embryologia* **10,** 318–322.
Uretz, R. B., and Zirkle, R. E. (1955). *Biol. Bull.* **109,** 370.
Uto, N., and Sugiyama, M. (1969). *Develop., Growth & Differentiation* **11,** 123–129.
Van't Hoff, J., and Kovacs, C. J. (1970). *Radiat. Res.* **44,** 700–712.
Walters, R. A., and Petersen, D. F. (1968). *Biophys. J.* **8,** 1487–1504.
Wells, P. H., and Giese, A. C. (1950). *Biol. Bull.* **99,** 163–172.
Yamashita, H., Mori, K., and Miwa, M. (1939). *Gann* **33,** 117–121.
Zeitz, L., Ferguson, R., and Garfinkel, E. (1968). *Radiat. Res.* **34,** 200–208.
Zeuthen, E., ed. (1964). "Synchrony in Cell Division and Growth." Wiley (Interscience), New York.
Zimmerman, A. M., and Zimmerman, S. (1967). *J. Cell Biol.* **34,** 483–488.
Zirkle, R. E. (1957). *Advan. Biol. Med. Phys.* **5,** 103–146.

Chapter 6

The Effects of Radiation on Mammalian Cells

CARL J. WUST, W. STUART RIGGSBY, and GARY L. WHITSON

I. Introduction

The radiosensitivity of mammalian cells, in spite of extensive research, is still one of the least understood areas in biology today. Almost without exception, the magnitude of difficulty in solving this area of research lies in our failure to understand normal processes and their regulation in mammalian cells. Consequently this has led to two main points of view concerning radiation studies. One view adopts the idea that one must first have a thorough knowledge concerning normal events in cells before defining what mechanisms (if any) are interrupted or invoked by various kinds and amounts of radiations. The other view asserts that the use of radiations, both ionizing and nonionizing, in the interfering with normal cellular events, will lead to a more definitive understanding of these normal cell processes. This chapter deals primarily with the latter view and evaluates questions concerning the nature of:

1. The control of cell division
2. Differences in the radiosensitivity of various phases of the cell cycle
3. The role of chromosome aberrations in mutation
4. Target molecules responsible for radiation lethality
5. The basis for radiosensitivity of the immune response
6. The molecular basis of radiation-induced cancer

The development and improvement of mammalian cell culture techniques during the past 20 years have provided radiation biologists with ample material for *in vitro* studies (for review, see Elkind and Whitmore, 1967). More recent discoveries on synchronization of mammalian cells *in vitro* by techniques such as the addition and removal of excess thymidine (Xeros, 1962), the collection of cells at metaphase by detachment (Terasima and Tolmach, 1963a), and the use of the drug Colcemid (Puck and Steffen, 1963) have now provided researchers with the opportunity to explore cellular radiobiological phenomena during specific phases of the cell cycle. Walters and Petersen (1968a,b) have recently begun to analyze variations in the radiosensitivity of the cell cycle. *In vivo* techniques for the study of cell renewal systems and analysis of the immune response in mammalian cells have also provided opportunities for the study of radiobiological phenomena. This chapter reviews some of the important findings concerning the radiosensitivity of mammalian cells both *in vivo* and *in vitro*. It is hoped that our presentation will suggest new ideas concerning radiation research on mammalian cell systems.

II. The Normal Cell Cycle

Until the early 1950's the concept of the cell cycle was limited to include only a long interphase and mitosis. Studies by Howard and Pelc (1951), using the technique of autoradiography, led to the discovery that the interphase or so-called "resting" phase constituted a very active period of biochemical synthesis, in particular, the synthesis of DNA. This led to the discovery (Howard and Pelc, 1953) that the interphase in certain plant cells was composed of a period (designated S) during which DNA of the cell replicated and that this period is preceded by a "gap" (designated G_1) and followed by a second "gap" (designated G_2). It was subsequently demonstrated by Lajtha *et al.* (1954) that these periods, G_1, S, and G_2, exist in mammalian cells. This generality has led to the common use of the diagram in Fig. 1, depicting the cell life cycle of mammalian cells.

Generation times for mammalian cells during the log phase of growth vary considerably, depending both on the species and the type of tissues from which the cells are derived (see Table I). In general, mitosis represents varying intervals for G_1, S, and G_2. Nondividing cells in living tissues such as liver, kidney, and small lymphocytes of the blood are considered by Lajtha (1968) to be in a state he refers to as G_0. This state is thought of as a kind of "stalled" G_1, since, if such cells are stimulated to divide, they first go through a short cycle. Gelfant (1962) has also identified populations of mouse epidermal cells *in vivo* which he calls "G_2 cells" because they are stalled or prolonged in G_2 for long periods of time before proceeding to cell division. Our concern here, however, is with cells progressing through the normal cell cycle with no particular attention to such nondividing differentiated cells.

There are few identifiable markers on the cell cycle map of mammalian

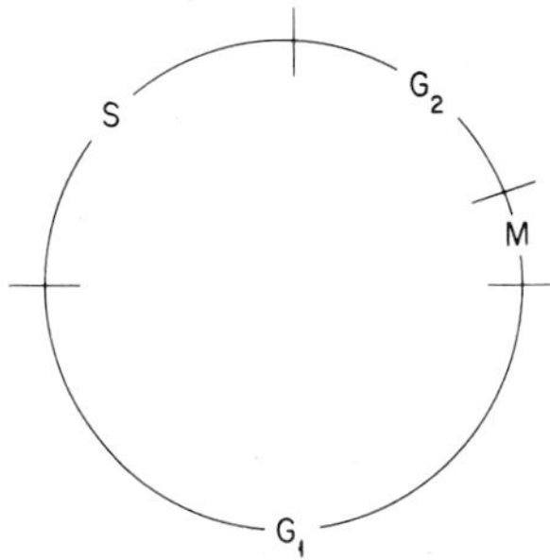

Fig. 1. The generalized life cycle of a mammalian cell.

TABLE I

Duration of Phases in the Cell Cycle of Log Phase Mammalian Cells

Cell line	Incubation temperature (°C)	G_1 (hr)	S (hr)	G_2 (hr)	M (hr)	Generation time (hr)	Reference
				In vitro			
Mouse fibroblast L-P59	37	9.1	9.9	2.3	0.7	22	Dewey and Humphrey (1962)
Mouse leukemia L5178Y		1.8	7.3	1.2	0.55	10.85	Watanabe and Okada (1967)
Chinese hamster fibroblast B 14-FAF28	37	2.7	5.8	21	0.4	11	Hsu *et al.* (1962)
Chinese hamster ovary		4.7	4.1	2.8	0.8	12.4	Puck *et al.* (1964)
Human carcinoma HeLa S3		8.4	6.0	4.6	1.1	10.1	Puck and Steffen (1963)
Human kidney		14	8	5	0.8	27.8	Galavazi and Bootsma (1966)
				In vivo			
Mouse Ehrlich ascites tumor		5.7	8.5	3.8	1	19	Kim and Evans (1964)
Mouse hair follicle		3	7	1.5	0.5	12	Cattaneo *et al.* (1961)

cells. The markers which have been found fall into three main categories: (1) structural or morphological, (2) biochemical or synthetic, and (3) physiological. Structural markers and their time of appearance during the cell cycle are shown in Table II. These events at mitosis (chromosomal behavior) are perhaps the most striking and most easily discernible by light microscopy methods. Changes in other structures such as centrioles, cytoplasmic organelles, and smaller structures like ribosomes are more easily detected by electron microscopy and biophysical methods. Biosynthetic markers are shown in Table III. This table emphasizes the fact that both the G_1 and G_2 are phases that involve essential biosynthetic events. For example, RNA and protein synthesis are initiated at the onest of G_1 and continue until the onset of mitosis. It also appears (Koch and Stokstrad, 1967) that mitochondrial DNA synthesis in human liver (Chang cell line) occurs during the G_2 period in the cell cycle. The scant data available on this subject, however, do not allow us to conclude that this might be the case in all mammalian cells. There are also some findings concerning physiological markers in the cycle map of mammalian cells (see Section III). Physiological markers represent points

TABLE II
Structural Changes During the Normal Cell Cycle of Mammalian Cells

Structure	Phase of formation and duration	Reference
Nuclear membrane	Appears at the initiation of G_1, remains until onset of mitosis	Robbins and Gonatas (1964)
Golgi apparatus	Appears during early G_1, remains until onset of mitosis	Robbins and Gonatas (1964)
Nucleoli	Appear during early G_1, disappear during early prophase	Elkind and Whitmore (1967)
Centrioles	Reproduce during S, migrate to poles during G_2	Robbins *et al.* (1968)
Spindle fibers	Form during G_2, Breakdown at late telophase of M	Robbins *et al.* (1968)
Polysomes[a]	Form during early G_1, Disaggregate at metaphase of M	Scharff and Robbins (1966); Steward *et al.* (1968)
Chromosomes	Splitting (conformational change for DNA synthesis); begins at initiation of S	Hsu *et al.* (1962); Dewey *et al.* (1966)
	Chromosomal duplication, entire S phase	Lajtha *et al.* (1954); Terasima and Tolmach (1963a)
	Coiling (condensation) begins in late G_2; uncoiling (decondensation) begins at initiation of G_1	Johnson and Holland (1965); Blondel and Tolmach (1965)
	Changes during mitosis	Elkind and Whitmore (1967)
	Prophase: Thicken (contract), appear rodlike, both chromatids of each chromosome apparent, joined by a single centromere	
	Metaphase: Chromosomes arranged on equatorial plate; most contracted state, spindle fibers attach to centromere from each centriole	
	Anaphase: Each centromere splits, one chromatid of each half drawn toward its respective centriole	
	Telophase: Chromosomes (chromatids) at opposite halves of cell begin to elongate and become threadlike	

[a] Considered as a structural change because they are an assemblage of ribosomes attached to mRNA and not treated as synthesis. Disaggregation probably reflects breakdown of mRNA.

TABLE III
Synthetic Events during the Cell Cycle of Cultured Mammalian Cells

Event	Phase of initiation	Duration of event (G_1 | S | G_2 | M)	Selected references
Nuclear RNA	G_1	|←——→| (G_1–S)	Doida and Okada (1967)
rRNA and tRNA	G_1	|←———→| (G_1–G_2)	Terasima and Tolmach (1963a); Scharff and Robbins (1965); Prescott and Bender (1962); Watanabe and Okada (1968)
S-Stage proteins	Late G_1	|← –Unspecified	E. W. Taylor (1965); Terasima and Yasukawa (1966)
DNA (nuclear)	S	|↔| (S)	Lajtha *et al.* (1954); Terasima and Tolmach (1963a); Kasten and Strasser (1966)
Histone	S	|↔| (S)	Robbins and Borum (1967)
Nucleolar RNA	Early S	|←Unspecified	Kasten and Strasser (1966); Crippa (1966)
Mitochondrial DNA	G_2	|←→| (G_2)	Koch and Stokstrad (1967);
Mitotic mRNA	Early G_2	- -Unspecified→|	Tobey *et al.* (1966)
Mitotic protein	Late G_2	- -Unspecified - →|	Littlefield (1966); Brent *et al.* (1965); Stubblefield and Murphree (1968)
Protein	G_1	|←———→| (G_1–G_2)	Prescott and Bender (1962); Johnson and Holland (1965)

that can be characterized by the use of inhibitors. J. M. Mitchison (1969) describes a physiological marker as a "point in the cycle before which an inhibitor stops the subsequent division and after which it does not." In most cases inhibitors are specific drugs and/or radiations.

III. Radiosensitivity of the Cell Cycle

In the usual sense, sensitivity of cells to radiation can be divided into two main categories: (1) lethal effects, and (2) sublethal effects. Lethal effects, as the name implies, mean cell killing and have been the subject of many investigations concerning radiation effects on mammalian cells (for review see Bond *et al.*, 1965; Elkind and Whitmore, 1967). Certain lethal effects will also be discussed in Section V. Sublethal effects are usually detected as division delay or mitotic delay as a consequence of radiations. It has been found that delay in cellular division depends largely on the position (phase) of cells in the cell cycle at the time of irradiation (Sinclair, 1968; Okada, 1970, Chapter 6,B).

The cell cycle in mammalian cells is sensitive to sublethal doses of ionizing radiations. The progress of irradiated cells from one phase of the cell cycle to the next varies according to three main criteria. These are (1) the cell line studied, (2) the dose of radiation, and (3) the position of cells (phase) in the cell cycle at the time of irradiation. For a review see Table VI-2 in Okada (1970). In general, the sensitivity of the various phases of the cell cycle to ionizing radiations, proceeding from the least sensitive phase to the most sensitive phase, is in the following order:

$$M < G_1 < S < G_2$$

The G_2 period is the most sensitive phase and several considerations to explain this order of increasing sensitivity have been investigated.

First let us consider cells in the M phase of the cell cycle at the time of irradiation. These cells usually divide at a normal rate and proceed to G_1 without delay (Terasima and Tolmach, 1963b; Walters and Petersen, 1968a). This is the time during the cell cycle when all of the essential biosynthetic events have been completed (Table I), although chromosomal movements are maximal (Table II). Carlson (1967), however, observed that grasshopper neuroblast cells, when irradiated with UV during prophase, revert back to interphase cells. Carlson (1969) also reported a similar reversion phenomenon by X-radiation of mouse mammary tumor cells. Doida and Okada (1969), however, failed to observe such X-ray-induced reversions.

Although irradiation of cells during the M phase does not result in immediate delays in cell division, analysis of the first postirradiation cycle indicates that there are delays in the subsequent cell division (Terasima and Tolmach, 1963b; Froese, 1966; Yu and Sinclair, 1967). Rustad (1960) was one of the first to report that radiation damage to cells could be carried over into the next cell cycle and be expressed at later stages. This led him to formulate a concept of "stored" radiation damage. According to Terasima and Tolmach (1963b), irradiation of HeLa S_3 cells with 300 rads of X-rays during the M phase results in a prolongation and depressed rate of DNA synthesis during S and a prolonged G_2 period. Others have reported similar findings in different cell lines, and it may be that storage of radiation damage with delayed effects is a widely occurring phenomenon. It is not known, however, what mechanisms or processes are involved in this phenomenon.

Progress of cells through G_1 to S can be estimated by continuous labeling of asynchronous populations with tritiated thymidine and sampling populations at intervals thereafter to determine by autoradiography the number of cells entering into S (Howard and Pelc, 1953; Stanners and Till, 1960). The phases of synchronized cells can be determined by pulse labeling (Terasima and Tolmach, 1963a).

Irradiation of G_1 cells *in vivo* (slowly dividing or nondividing cells) results in a "G_1 depression" (Lajtha *et al.*, 1958a) or delay of entry of cells into S. On the other hand, irradiation of rapidly dividing cells (cancer cells both *in vivo* and *in vitro*) does not affect or delay cell progress through G_1 to S (Yamada and Puck, 1961; Whitmore *et al.*, 1961; Kim and Evans, 1964; Watanabe and Okada, 1966).

The precise mechanism of G_1 depression is not clearly understood, but it is known that irradiation during this period in regenerating liver results in both a depression of nuclear RNA and protein synthesis (Sarkar *et al.*, 1961; Welling and Cohen, 1960). There is evidence that irradiation after this period does not affect total RNA or protein synthesis (Puck and Steffen, 1963; Van Lancker, 1960). However, as shown in Table I, a mitotic mRNA and specific mitotic proteins are synthesized during G_2. It has been suggested that irradiation in G_2 interferes with these syntheses (Walters and Petersen, 1968a). Before cells can traverse from G_1 to S, enzymes necessary for DNA replication must be synthesized (Okada and Hempelmann, 1959; Bollum *et al.*, 1960), and these enzymes appear to be sensitive to irradiation. In addition, Lehnert and Okada (1963, 1966) have shown that the structure of deoxyribonucleoprotein is altered by irradiation. Thus DNA synthesis is delayed as well as entry of cells into S.

It is interesting that irradiation of cells in G_1 delays the first postirradiation cell cycle less than irradiation of any of the other phases in the cell cycle (Terasima and Tolmach, 1963b). In other words less radiation damage is "stored" and expressed at later stages in cells irradiated in G_1. This may be an indication of the activity of highly efficient repair mechanisms or the absence at this time of whatever it is that stores radiation damage in cells.

The S phase of several mammalian cell lines is very sensitive to ionizing radiations (Painter and Hughes, 1961; Mak and Till, 1963; Terasima and Tolmach, 1963b; Watanabe and Okada, 1966). This sensitivity usually results in a depression in the rate of DNA synthesis and the accumulation of cells in a prolonged S phase. This phenomenon was first referred to by Painter and Robertson (1959) as the "S retention effect."

The mechanism of depression of DNA synthesis by irradiation is open to speculation. Ord and Stocken (1958) suggest damage to the DNA template, while Lajtha *et al.* (1958b) proposed damage to enzymes required for DNA synthesis as responsible for radiation-induced depression. Although the rate of DNA synthesis is depressed in irradiated S, most cells eventually are able to synthesize a full complement of DNA. Indeed in some cases, excess DNA is synthesized following irradiation (Smets and Dewaide, 1966; Watanabe and Okada, 1968). It has been suggested that such excess DNA synthesis represents "repair synthesis" (Okada, 1970, Chapter VI,D) or "unscheduled synthesis" (Rasmussen and Painter, 1966). However, in the usual sense, as discussed in Chapter 4, repair synthesis should result in no net gain in the amount of DNA as compared with normal semiconservative DNA replication where there is a limited net gain (doubling) in the amount of total DNA in each cell. It seems, therefore, that there is no satisfactory explanation at this time for the synthesis of excess DNA as a result of irradiation. Several hypotheses are presented by Okada, (1970, Chapter VI,C) in an attempt to explain the accumulation of S phase cells and the mechanisms of depression of the rate of DNA synthesis. However, there are still no convincing proposals for this problem.

It is also interesting to note that irradiation of HeLa cells in S not only prolongs entry of cells into G_2, but a prolonged G_2 results in division delays during the first postirradiation cell cycle (Terasima and Tolmach, 1963b).

The G_2 period of the mammalian cell cycle is the most sensitive period (produces the longest mitotic delay) to ionizing radiations. This delay of entry of cells into mitosis has been termed the "G_2 block" (Terasima and Tolmach, 1963b; Puck and Steffen, 1963). The location of the G_2 block has been determined by a variety of methods and appears almost certain to occur in either mid or late G_2 (Puck and Steffen, 1963; Walters and Petersen, 1968a,b; Doida and Okada, 1969). The mechanism(s) of the G_2 block has been the subject of much discussion. The best explanation relates the finding of a G_2 block coincident with inhibited synthesis of a specific mitotic protein (see Table I and Walters and Petersen, 1968a). There is, however, strong support for the hypothesis that the production and repair of chromosome aberrations are the cause for the G_2 block (Puck and Yamada, 1962; Puck and Steffen, 1963).

In summary, it can be concluded that all phases of the cell cycle in mammalian cells are sensitive to ionizing radiations. The most sensitive phase is G_2, just prior to the onset of mitosis. There is some indication that the synthesis of mitotic protein and its inhibition by irradiation is responsible for the longest or maximal delays in cell division.

IV. Chromosome Aberrations

One of the most dramatic effects of radiation at the cellular level is the production of chromosome aberrations. These aberrations are observable only at the M stage of the cell cycle and therefore are limited to light microscopy study. Information will be presented here concerning the types of chromosome aberrations that are formed in relation to the time at which cells are irradiated in the cell cycle. Irradiation of cells in G_1 results in a preponderance of chromosome-type as well as some chromatid-type aberrations. If irradiation takes place during early S phase, chromatid aberrations far outnumber the chromosomal type. Only chromatid-type aberrations are seen in cells irradiated at mid S phase through G_2. Aberrations are significant as markers in that they occur in greatest frequency in cells and tissues that are highly proliferative or rapidly dividing (Bender, 1969).

Although there have been extensive reviews on radiation cytogenetics and the production of chromosome aberrations (Evans, 1962; Bender, 1969), the dangers to the well-being of organisms or cells containing chromosome aberrations is still not well understood. Bender (1969), however, from a cytogeneticist's point of view has said that "since the aberrations represent changes in the cells' genetic information, it is very easy to believe that they [the aberrations] should have profound effects on their [the cells'] ability to survive and function." Thus, on the one hand, somatic chromosome aberrations are potential hazards to the individual, whereas germ cell aberrations transmitted through successive generations could result in abnormalities in future progeny.

Aberrations produced by radiations that usually have a low linear energy transfer (LET) result from chromosome breaks. Such radiations are produced by hard X-rays or γ-rays. High-energy LET radiations such as α-particles, fission neutrons, or low-energy protons also produce chromosome breaks but not with the same dose–effect kinetics as do radiations of low LET. Dose–effect kinetics for the production of chromosome breaks have been covered in detail by Lea (1955), Elkind and Whitmore (1967), and Bender (1969).

The kinds of breaks observed in mammalian cells fall into two major classes, those that are of the chromosome type and those that are of the chromatid type. Chromosome breaks occur in cells before DNA replication takes place at a time when chromosomes are actually single chromatids. Typical lesions (breaks) that lead to chromosome type aberrations are shown in Fig. 2. There are five basic types of these aberrations: deletions, inversions, rings, translocations, and dicentrics. Of these basic types, translocations and dicentrics involve chromosome exchanges. The frequency of aberrations in Chinese hamster cells irradiated at different phases in the cell cycle suggests that G_2 cells have the highest frequency and S cells the lowest frequency (Yu and Sinclair, 1967). Chromatid aberrations appear most frequently in G_2 cells in human leukocytes in culture and least frequently in G_1 cells (Brewen, 1965). The types and frequencies of human chromosome aberrations (Chu *et al.*, 1961; Bender, 1969) and those of other mammalian cells both *in vivo* and *in vitro* (Elkind and Whitmore, 1967) have been the subject of extensive review.

One of the most interesting aspects of the study of chromosome breaks concerns the underlying mechanism of the restitution process (repair) or rejoining between broken ends of chromosomes that results in aberrations. It is, in fact, these processes or their failure that result in chromosome exchanges, persistence or repair of single breaks, and the formation of the various types of observed aberrations (for review, see Lea, 1955; Elkind and Whitmore, 1967; Casarett, 1968; Bender, 1969; Okada, 1970). Very little is known about these processes, particularly in mammalian

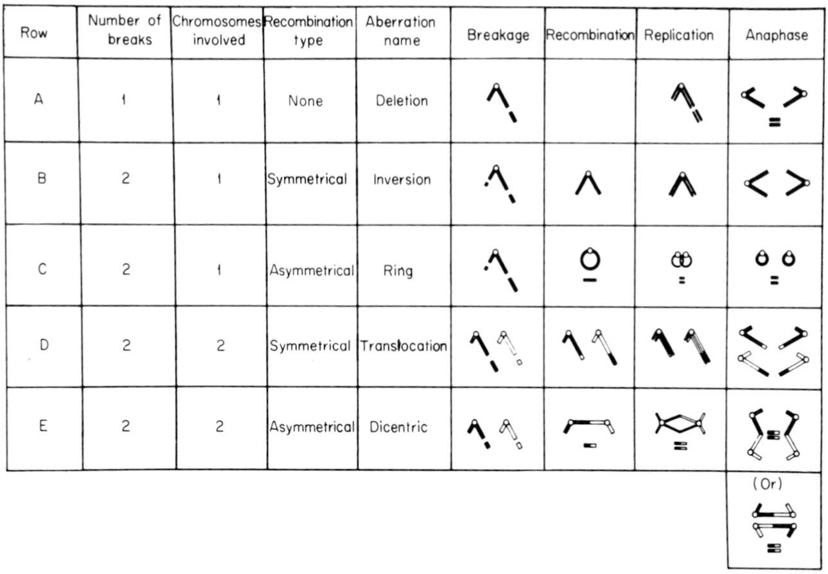

Row	Number of breaks	Chromosomes involved	Recombination type	Aberration name	Breakage	Recombination	Replication	Anaphase
A	1	1	None	Deletion				
B	2	1	Symmetrical	Inversion				
C	2	1	Asymmetrical	Ring				
D	2	2	Symmetrical	Translocation				
E	2	2	Asymmetrical	Dicentric				(Or)

Fig. 2. Schematic representation of the origins of various chromosome-type chromosomal aberrations. (From Bender, 1969, courtesy of the author.)

cells. Dewey and Humphrey (1964) have observed that Chinese hamster ovary cells are capable of very rapid repair of breaks during G_1 but that S phase cells are not. They further conclude that rejoining is favored over repair in the S and G_2 phases of the cell cycle. Virtually nothing is known about the chemistry of rejoining of chromosome breaks in mammalian cells, and there is little hope of clarifying this aspect of the problem until more is known about the details of mammalian DNA replication in general. There have, however, been reports concerning certain biochemical events associated with restitution and rejoining processes in *Vicia faba* (bean root-tip cell) chromosomes (Wolff and Luippold, 1955). These investigators demonstrated that ATP is necessary for both processes. Wolff (1960) has further demonstrated that protein synthesis is required. Evidence from mammalian cells indicates that oxygen enhances the number of initial chromosome breaks (Deschner and Gray, 1959; Valencia and DeLozzio, 1962) but this does not relate to the nature of restitution or rejoining processes. These are, admittedly, only preliminary thrusts in the direction of solving this very complicated problem.

The fate of chromosome aberrations *in vivo* has been followed in irradiated mouse bone-marrow cells (Nowell and Cole, 1963), and there is some indication that low dose rates of ionizing radiations result in a significant but low incidence of persistent chromosome aberrations over a long period of time. The persistence of chromosome aberrations *in vivo* in humans (Nowell, 1965, 1967) is difficult to analyze because of the combined use of drug treatment and therapeutic irradiation of the patients. Bender (1969) has reviewed this subject and stressed the need for improved methods of chromosome analysis as well as the need for large-scale population studies to assess the potential hazards of radiation on human chromosomes and human genetics. Clearly, much more information is required than is now available.

V. Radiation Lethality and Cell Renewal Systems

Cell renewal systems are systems in which cell populations are in steady-state equilibrium. This steady state is established when the number of cells entering a system equals the number of cells leaving it. Examples of mammalian cell renewal systems are: (a) cells that line the gastrointestinal tract; (b) the skin; (c) blood with its associated bone marrow hemopoietic tissues; and (d) lymphatic tissues. Lethal doses of ionizing radiations have profound effects upon the regenerative capacity

TABLE IV
Survival Times in Lethally Irradiated Selected Mammals

Species	Radiation dose and type	Approximate survival time (days)	Selected references
Mouse	1200 rads, 250 kV X-rays	3	Silverman *et al.* (1958)
	1500 rads, 1.2 MV-γ	3 1/2	Langham *et al.* (1956)
Rat	1500 rads, 250 kV X-rays	3 1/2	Sullivan *et al.* (1959)
	1000 rads, 1.2 MV-γ	4	Langham *et al.* (1956)
Dog	1800 rads, 2000 kV X-rays	3 1/3	Handford (1960)
	3000 rads, 3000 kV X-rays	3 3/4	Handford *et al.* (1961)
Monkey	800 rads, 250 kV X-rays	6	Allen *et al.* (1960)
	1000 rads, 1.2 MV-γ	6	Langham *et al.* (1956)
Sheep	518 rads, fission spectrum neutrons	10	Riggsby *et al.* (1966)
	476 rads, fission spectrum neutrons	15	Riggsby *et al.* (1966)
	476 rads, 250 kV X-rays	20	Riggsby *et al.* (1966)
	359 rads, 250 kV X-rays	28	Riggsby *et al.* (1966)
Man	$>$1000 rads, mixed fission products	6	Alexander (1957)
	700 rads, mixed fission products	50	Alexander (1957)

of cells that form these tissues. Delayed, as opposed to immediate, death of an organism as a consequence of ionizing radiation is often linked with the failure of cell renewal to operate in one or several of these cell systems. Survival times of certain mammals after ionizing radiations are shown in Table IV.

A. The GI Syndrome

The effects of radiation on the gastrointestinal tract is commonly referred to as the GI syndrome. This syndrome manifests itself in animals with whole-body exposures to radiation. It results in the eventual destruction of epithelial cells. Thus when lethality is observed following irradiation, it is usually expressed as failure of renewal of GI tract cells.

Changes in the epithelial lining of the GI tract of the small intestine of the mouse after total body irradiation result in pyknosis of nuclei, swelling of cells, and eventual lysis (Montagna and Wilson, 1955). Epithelial cells are very sensitive to X-irradiation and cellular destruction has been observed with doses as low as 100 rads in the mouse (Knowlton and Hempelmann, 1949), in the rat (Williams *et al.*, 1958), and in man (Tori and Gasbarrini, 1963). Some cellular regeneration is observed soon after sublethal exposures (100 rads) but lethality without regeneration usually occurs at doses of 2000 rads and higher (Quastler, 1956).

Cells that survive in the epithelial lining show considerable enlargement and lateration of their nuclei (McGrath and Congdon, 1959; McGrath, 1960). The most extensive cytological changes apparently are in the crypts of the intestine where the stem cells are believed to originate (Quastler and Hampton, 1962). The most pronounced effects that have been observed are abnormal mitoses (Montagna and Wilson, 1955; Lesher and Vogel, 1958), changes in the microvilli, alteration in the structure of mitochondria, and the disappearance of the endoplasmic reticulum (Quastler and Hampton, 1962; Detrick *et al.*, 1963). Radioisotopic labeling of crypt cells with tritiated thymidine indicates that DNA synthesis continues in abnormal cells as well as in those that are destined to die (Sherman and Quastler, 1960). This may be an indication that G_2 cells in the stem cell compartment are more sensitive to ionizing radiation and that other phases of the cell cycle such as M and G_1 are not impaired from entering S. It is not known, however, whether this is the case or whether cells synthesizing DNA under these conditions have completed DNA synthesis. It may be that cells that die do not complete S and that unscheduled DNA synthesis is occurring in an attempt at repair. It has been observed that mouse crypt cells that survive 2500 rads of X-rays triple their nuclear size within 30 hr following irradiation (McGrath and Congdon, 1959). There is no information as to whether these abnormally large cells are the result of cellular fusion or of an increase in synthetic activity. Perhaps these are cells that skip phases of the cell cycle, e.g., progress from G_2 directly into G_1 with no intervening mitosis or cell division.

B. The Bone-Marrow Syndrome

The stem-cell compartments of circulating granulocytes in the blood are very sensitive to irradiation and their destruction is linked with the manifestation of the GI syndrome. The leukocytes play an important role by engulfing bacteria and foreign substances that invade the bloodstream and tissues in mammals. One of the most significant changes in animals after lethal doses of ionizing radiation is the depletion in numbers of these circulating granulocytes. This observation strongly infers that the stem cells in the bone marrow that form these elements are either destroyed or have had their proliferation ability impaired. The

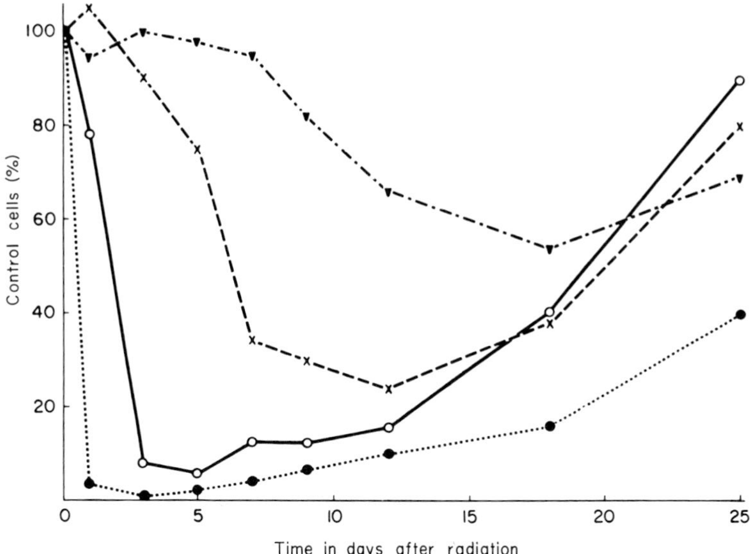

Fig. 3. Typical change in the numbers of circulating blood cells of rats at various times after exposure to a moderate dose of radiation (500–1000 rads). ▼, erythrocytes; ×, platelets; ○, granulocytes; ●, lymphocytes. (From Casarett, 1968, courtesy of the author and the publisher.)

number of circulating granulocytes is almost zero at the time the GI syndrome appears. Changes in granulocyte counts following total-body X-radiation are shown in Fig. 3.

Rytömaa and Kiviniemi (1968a,b,c) have reported two specific humoral regulators (chalone and antichalone) that control granulocyte production and are produced somewhere in cells of the hemopoietic system. They theorize that some sort of feedback regulation by chalone and antichalone control the levels of granulocytes in the blood and that chalone stimulates production, whereas antichalone turns off production. These investigators further believe that the radiosensitivity of granulocyte production is directly related to the destruction of these humoral regulators or an interference with the synthesis of these substances. This is an attractive hypothesis, but more studies are needed to prove the existence of such substances and their origin in the bone-marrow system.

Early in the history of radiobiology it became apparent that the hemopoietic system was one of the most radiosensitive systems in the body and that death in animals as a result of low doses of ionizing radiation could usually be attributed to failure of cell renewal in the blood (for review, see Congdon, 1959, 1971; L. H. Smith and Congdon, 1960).

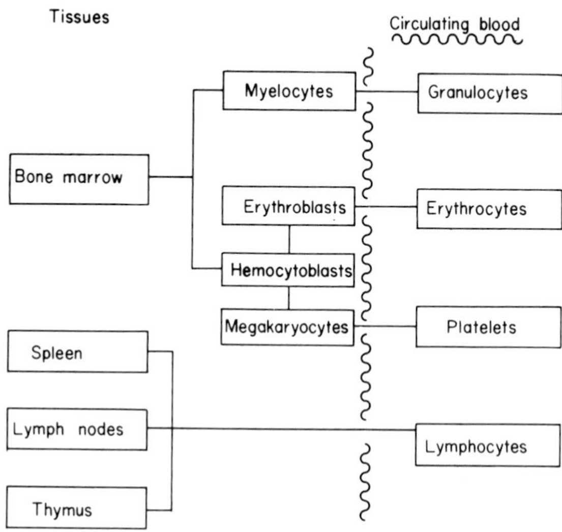

Fig. 4. Schematic representation of the origins of circulating blood cells from the stem-cell compartments of bone marrow and related hemopoietic tissues.

Changes in cell composition and cell numbers in the hemopoietic system after acute radiation exposure in mammals has been named the bone-marrow syndrome. Tissues usually involved in hemopoietic cell death are bone marrow, thymus, lymph nodes, and spleen. The origins of the cell types and the interrelationship of the various stem cell-producing tissues of the hemopoietic system involved in the production of circulating blood cells are shown in Fig. 4.

Within the last decade a large amount of information has been collected concerning radiation effects on bone-marrow-derived cells. Most of the recent findings have been reviewed by Bond *et al.* (1965), Elkind and Whitmore (1967), and Congdon (1971). The present discussion is concerned mainly with the variations in radiosensitivity of cell renewal populations of the blood, bone marrow, spleen, and lymphatic tissues.

1. *Circulating Blood Cells*

The various cellular components of circulating blood show different sensitivity to ionizing radiations. Whole-body irradiation at moderate doses (500–1000 rads) will lead eventually to a decrease in the numbers of all circulating blood-cell types (Fig. 3). This decrease in numbers and types of cells can result from any one of or a combination of the following conditions: (a) direct destruction; (b) hemorrhage or leakage through capillaries; (c) damage to stem cells with accompanying loss of replication. As shown in Fig. 3, lymphocytes are the most radiosensitive and have the lowest rate of recovery; mature granulocytes, on the other hand, are radioresistant, but their life-span in circulating blood has been estimated to be only about 4 to 5 days (Bond *et al.*, 1965) so that the loss in granulocytes following irradiation is most likely due to condition (c) above. Blood platelets, which play an important role in clotting, reach their lowest levels in the blood about 2 weeks after irradiation. These elements arise from megakaryocytes in the bone marrow and represent a depression in proliferation of these progenitor cells at this time. The period during which there are low levels of platelets in the blood is believed to contribute to the characteristic hemorrhaging and subsequent loss of clotting ability that is one of the typical symptoms of the bone-marrow syndrome (for more detail see Casarett, 1968).

2. *Bone-Marrow Stem Cells*

The stem-cell compartments of the bone marrow are among the most radiosensitive cells in the hemopoietic system. Other cell types of the bone marrow, such as fat cells, macrophages, reticular cells, and connective tissue are resistant to doses of radiation that normally elicit the bone-marrow syndrome (Casarett, 1968). Histological changes in bone marrow following acute radiation exposures usually results in the proliferation of fat-like tissues, a decrease in blood content, and a drastic reduction in the number of nucleated cells (for review see Bond *et al.*, 1965). The relative magnitude of sensitivity of bone-marrow cells in the rat after a moderate dose of ionizing radiation (500–1000 rads) is shown in Table V. Erythroblasts are the most sensitive of the stem cells.

TABLE V

Summary of Changes in Hemopoietic Cells of Rat Bone Marrow following a Moderate Dose of Radiation[a]

Phase	Erythroblasts	Myelocytes	Megakaryocytes
Detectable decrease in numbers	3 hr	2 days	3 days
Minimum numbers	4–5 days	4–5 days	4–13 days
Recovery	7–8 days	12–40 days	14–41 days

[a] From Casarett (1968), courtesy of the author and the publisher.

3. *Spleen Cells*

Spleen cells, particularly those that comprise the germinal centers of the white pulp, are sensitive to ionizing radiations. Large doses of whole-body radiation usually result in a complete destruction of the lymphocytic centers in the spleen, leading to atrophied, "sterilized," spleens. The possibility of rescuing mice receiving supralethal doses of ionizing radiation by transplantation of isologous bone marrow (IBM) cells from normal mice led to the discovery that new splenic nodules can arise in spleens that were sterilized by irradiation (McCulloch and Till, 1960; Till and McCulloch, 1961). These investigators found that intravenously injected IBM cells migrate through the circulating blood, proliferate, and form nodules of lymphocytic cells in radiation-sterilized spleens. In other words, sterilized spleens act as *in vivo* culture chambers. It was later determined that the number of splenic nodules formed was directly proportional to the number of IBM cells transferred into the mouse, indicating that one bone-marrow stem cell yielded one spleen colony (Till and McCulloch, 1964).

The relationship between the number of intravenously injected nucleated marrow cells and the number of colonies observed in preirradiated spleens is shown in Fig. 5. Note that the data refer to colony-forming units (CFU) and not necessarily to single cells.

The obvious limitation in the Till and McCulloch method has been the failure to identify the progenitor(s) of the spleen colonies that are formed

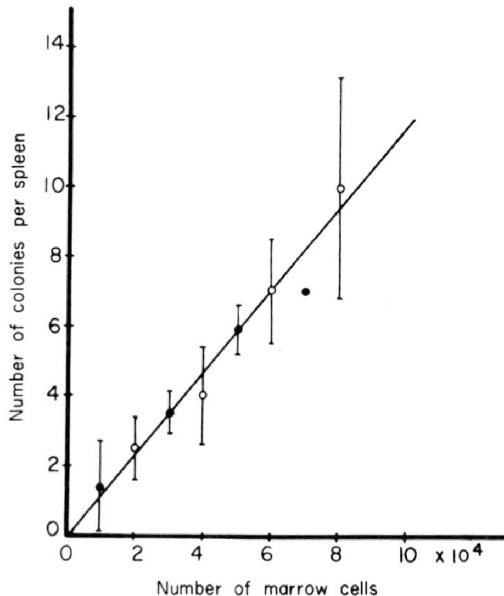

Fig. 5. The relationship between the number of nucleated marrow cells injected and the mean number of colonies formed in the spleen. The results of two separate experiments showing the mean and standard deviation for each point. (From Till and McCulloch, 1961, courtesy of the authors.)

by IBM injections, particularly since a bone-marrow suspension must necessarily contain many cell types. Other experiments must be devised in order to clarify this problem.

4. *Lymph Nodes and Thymus Cells*

Lymph nodes and thymus are relatively resistant to ionizing radiation (Casarett, 1968). When the destruction of cells in these tissues occurs, as in the spleen, this results from the direct effects of radiation on the reproduction of lymphocytes. Structural cells in these organs do not appear to change, and regeneration of lymphocytes usually occurs within a few days following moderate radiation doses (500–1000 rads). A summary of these changes is shown in Table VI.

TABLE VI
Summary of Changes in Irradiated Lymphatic Organs of Rats following Moderate Radiation Doses[a]

Phase	Lymph nodes	Thymus	Spleen
Destructive	1–7 hr	1–8 hr	1–8 hr
Debris removal	7–24 hr	8 hr to 2 days	8–24 hr
Inactivity	1–2 days	2–9 days	1–14 days
Regeneration	2–25 days	10–28 days	14–49 days

[a] From Casarett (1968), courtesy of the author and the publisher.

VI. Radiosensitivity of the Immune Response

A. The Immune Response

Before considering the effects of irradiation on the immune response, it is necessary to briefly analyze the response. There are many comprehensive reviews available and no attempt will be made here to discuss all possible aspects of the response in detail (see Abramoff and LaVia, 1970).

The immune response is the sequence of events that occurs as the result of the exposure of an animal to a foreign, nonself-substance. This sequence involves the interaction of several different cell types that eventually results in the proliferation and differentiation of those cells that either synthesize immunoglobulins or react in a specific way to the foreign substance (cell-mediated immunity).

1. *Antigens*

Antigens are most often macromolecules that include proteins, polysaccharides, and nucleic acids. Although a macromolecular species is a necessity (it is usually a carrier protein), the antigenic determinant comprises a small portion of the molecule and has been shown to be a cluster of amino acids, oligonucleotides, or oligosaccharides that are of a size equivalent to about 1000 daltons (Kabat, 1968).

2. *Immunoglobulins (Rowe, 1970)*

Immunoglobulins are soluble proteins produced by plasma cells. They contain antibody-combining sites that react specifically with the antigenic determinant because of their complementarity, and these sites are comparable to the determinant in size. Five immunoglobulins have been described.

1. IgA, found primarily in secretions and body fluids and produced by local plasma cells. The molecule in secretions is a dimer, the two units of which are held together by a joining protein piece (Halpern and Koshland, 1970). In addition, there is a polypeptide secretory piece that is added by the epithelial cells during passage and secretion through these cells (Cebra, 1969). IgA does not activate complement.

2. IgM is a macroglobulin found in serum. It is the immunoglobulin synthesized early in the response, and its continued production is inhibited by the presence of the specific IgG (Jerne, 1967). It is regarded as the first line of defense, and the cells that produce it do not take part in long-lived synthesis or immunological memory (Uhr, 1964). Phylogenetically, IgM appears to be more primitive than other immunoglobulins (R. T. Smith *et al.*, 1966).

3. IgG is the γ-globulin class usually referred to as circulating antibody and is found after the injection of antigen. Plasma cells that produce IgG are able to synthesize antibody molecules for long periods of time, and their precursors participate in the development of a persisting immunological memory. The primary site of synthesis of IgM and IgG is in the spleen and lymph nodes.

4. IgD is a globulin that is found in low concentrations in blood; its biological function is unknown although it contains antibody-combining sites (Gleich *et al.*, 1969).

5. IgE is a homocytotrophic globulin found in low concentration in blood. It binds to mast cells and platelets where it mediates immediate-type hypersensitivity (Ishizaka and Ishizaka, 1968). IgG may be involved also in immediate-type hypersensitivity although release of the chemical mediators from the mast cells requires fixation and activation of the first five components of complement.

The fundamental unit of structure of immunoglobulins, regardless of their class, consists of two pairs of identical chain. The longer chains were designated heavy chains and the smaller chains light chains. The differences between classes of immunoglobulins are due entirely to differences between their heavy chains. The light chains are common to all immunoglobulin classes but occur in two forms or types. These are designated type K or type L and the chains as κ and λ. Differences between these are in the primary amino acid sequences.

Further subdivisions of immunoglobulins are recognized. For example, subclasses occur in IgG, IgA, and IgM. In addition, allotypic specificities have been detected in human immunoglobulins including more than 20 Gm factors for γ (IgG) heavy chains, three Inv (κ light chain) factors, and the O_z (λ light chain).

The antibody-combining site is located in the N-terminal portions of

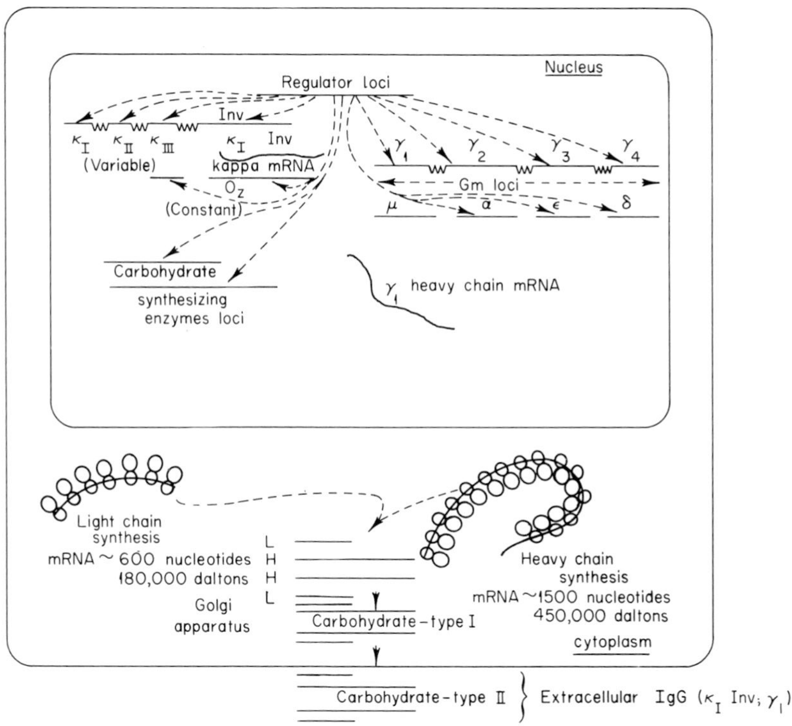

Fig. 6. Schematic representation of synthesis and assembly of an immunoglobulin molecule in the plasma cell. Synthesis involves the concerted effort of several genes and the simultaneous gene exclusion of all other possibilities. The secreted immunoglobulin of a single plasma cell represents a homogeneous population of one type. The immunoglobulin accounts for 60% of the soluble protein synthesis of that cell and is secreted at a rate of 2000–5000 molecules/cell/sec.

the light and heavy chains in the region referred to as variable. The complexity of assembly and concurrent gene exclusion is represented in Fig. 6. (For more detail see Gally and Edelman, 1970; Millstein and Monro, 1970.)

3. *Interaction of Cell Types*

Immunoglobulin biosynthesis or cell-mediated immunity depends upon the interaction of several cell types including cells derived from thymus, bone marrow, and the bursa-equivalent (see Fig. 7). In fowl, removal of the bursa of Fabricius results in the inability of the bird to synthesize immunoglobulins (Glick, 1970). The bursa-equivalent in mammals is believed to involve the tonsils, appendix, Peyer's patches, and/or the sacculus rotundus (Cooper *et al.*, 1966; Abramoff and LaVia, 1970).

Hormonal-like substances can also be secreted by the thymus (Goldstein *et al.*, 1966, 1967) and the bursa of Fabricius (St. Pierre and Ackerman, 1965) that affect the immune response.

a. Bone Marrow. Bone-marrow stem cells are derived from yolk sac and are influenced in development by the bursa or bursa-equivalent. These stem cells are responsible for populating the spleen, lymph nodes, and other tissues with macrophages (adherent to glass and plastic) and

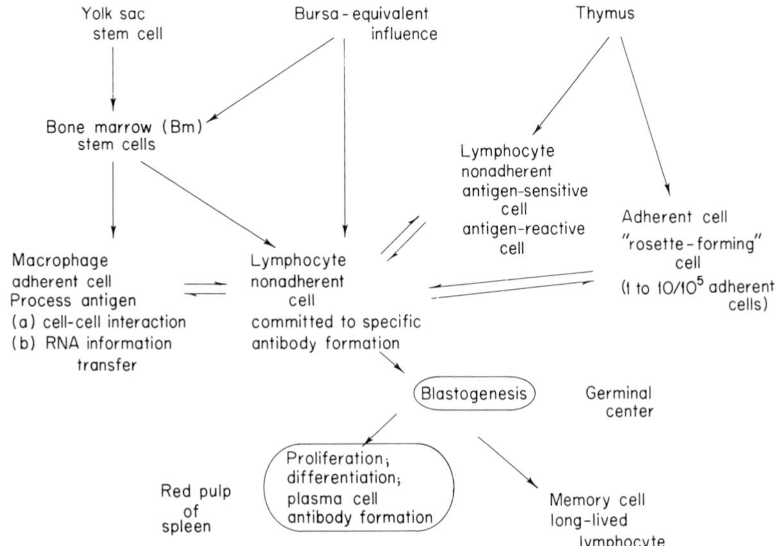

Fig. 7. Cellular interactions in the immune response. Single arrows indicate development or influence; double arrows indicate cell–cell interactions.

lymphocytes (nonadherent) (R. B. Taylor, 1969; Mitchell and Miller, 1968). The latter type develop into plasma cells after stimulation and are referred to as X cells (Sercarz and Coons, 1962) or the committed precursors of the antibody-forming cell (Abdou and Richter, 1970). Although no antibody is made in the bursa or bursa-equivalent, the removal of the tissue prenatally results in the loss of the capacity of the animal to synthesize immunoglobulins. It is probable then that bone marrow-derived lymphocytes must either pass through the bursa-equivalent, or the bursa-equivalent must influence the cells that develop into plasma cells.

b. Thymus. Although the thymus is primarily involved in cell-mediated immunity, it has been shown that a nonadherent, thymus-derived lymphocyte and an adherent, thymus-derived cell are necessary for maximal efficiency in development of plasma cells (Mosier *et al.*, 1970). The thymus then establishes a cellular basis for a normally functioning immune system, and this function may be restricted to a short period following development of the organ itself. The development of the immune potential, the restoration of this potential after damage, and maintenance of the potential are functions of the thymus (Miller, 1964).

Adherent cells from mouse spleen contain a subpopulation of 1–10/10,000 cells that are essential for the *in vitro* response to sheep erythrocytes (Mosier and Coppleson, 1968). It has also been demonstrated that a subpopulation of adherent cells interacts with erythrocyte antigens to form a "rosette." Rosettes are formed by the rapid and dense clustering of heterologous erythrocytes on the surface of about 1/10,000 adherent cells. Rosette-forming cells are different from plaque-forming cells and are apparently specific for different noncross-reacting heterologous erythrocytes. It is tempting to speculate that this essential subpopulation of adherent cells plays a role in concentrating antigen and that it is thymus-derived (Roseman, 1969).

That adherent and nonadherent, thymus-derived cells are required for immunoglobulin production is based on the loss of efficiency in *in vitro* experiments in which plaque-forming cells are observed. Plaque-forming cells are IgM- or IgG-producing cells in which the antigen is the erythrocyte. When the animal is thymectomized either prenatally or neonatally, the number of plaque-forming cells in greatly reduced unless the thymus is reconstituted (Mosier *et al.*, 1970). Cells derived from the thymus can be identified in immunoglobulin-synthesizing tissue by their reactivity with anti-θ serum. The θ antigen is present predominantly on thymus cells and in brain tissue (Reif and Allen, 1966) and is used as a cell marker for thymus-derived or thymus-dependent cells, distinguishing them from cells not previously influenced by the thymus (Chan *et al.*, 1970). It has been postulated that thymus-derived cell populations serve as a device for the antigen to reach cells with low affinity receptors (N. A. Mitchison, 1969).

4. *Antigen Processing*

Macrophages are found throughout the body as part of the reticulo-endothelial system (RES). Although these cells apparently function in a similar fashion, they can be subclassified on the basis of location and morphology; for example, peritoneal, alveolar, Kupffer (liver), microglia, and monocytes of the blood. They comprise in some instances a major cellular component of a tissue as in the spleen and liver. Kupffer cells constitute approximately 30% of the nucleated cells of the liver (Pearsall and Weiser, 1970).

Macrophages ingest and sequester soluble antigen within pinolysosomes. Particulate material is phagocytized and found in the phagolysosome. Either antigenic type may be completely degraded, partially degraded, and sequestered, or processed in one of several ways that would facilitate the immune response. As noted above, the majority of macrophages in the RES are derived from bone marrow.

Processing of antigen by the macrophage may involve:

1. A cell–cell interaction between macrophage and lymphocytes resulting in the antigen or antigenic determinant being presented to the lymphocyte in such a way that the lymphocyte is activated to proliferate and differentiate into a plasma cell (N. A. Mitchison, 1969). It appears that the immunocompetent, nonadherent cell is the lymphocyte in which blastogenesis and differentiation can occur after appropriate stimulation.

2. The synthesis and/or formation of an RNA-immunogen complex (RNA being approximately 23–28 S) that can be transmitted to the immunocompetent cell (Gottlieb *et al.*, 1967; see also Abramoff and LaVia, 1970). This complex loses its immunogenicity either by treatment with ribonuclease or by treatment with pronase, which destroys the protein moiety.

3. The synthesis and/or formation of an RNA (8–12 S) that is released and is informational to the receptor immunocompetent cell. This RNA was shown to provide information for the subsequent production of specific antibody both *in vitro* (Bell and Dray, 1969) and *in vivo* (Bell and Dray, 1971). In addition, immunoglobulins produced in recipient cells is of the allotype of the macrophage or processing cell (Bell and Dray, 1971). In these latter experiments, 30% of the cells in the recipient rabbit were found to be synthesizing immunoglobulin of the donor's allotype.

Delayed sensitivity skin reaction has been conferred on nonsensitive guinea pigs with RNA extracted from lymph nodes of sensitive animals (Jureziz *et al.*, 1970).

Although informational RNA can now be readily isolated in several laboratories, there has been hesitation in concluding that this material is informational since it is not understood how the information contributes to the receptor cell. One possibility presents itself in the finding that RNA-dependent DNA polymerase (reverse transcriptase) is possibly ubiquitous in mammalian cells (Scolnick *et al.*, 1971). This suggests that the informational RNA can be transcribed into DNA which could then be integrated into the genome of the recipient cell. In this context, it is not necessary that the RNA contain the total information for the entire amino acid sequence of both heavy- and light-chain polypeptides. The antibody-combining site consists of amino acid sequences of no more than 20 residues (Wu and Kabat, 1970) in the variable region and differences in the allotypes are 12 to 15 residues (Bell and Dray, 1971).

An alternative hypothesis can be constructed using the regulatory scheme suggested for higher organisms by Britten and Davidson (1969). They theorize that "producer" genes (e.g., genes that code for messenger RNA) are regulated by "activator" RNA molecules that are synthesized on "integrator" genes. The effect of the integrator genes is to induce transcription of many producer genes in response to a single molecular event. Thus informational RNA obtained from macrophages after processing could in effect be activator RNA in this mode. Activator RNA would complex with regulator genes that influence genes specifying antibody production in a single cell (Fig. 7). If all the immunoglobulin-producing genes were under the negative control of regulators, one might suppose that a stimulator-type fragment of RNA, transcribed in the macrophage, could result in the activation of specific structural genes for immunoglobulin synthesis.

5. *Blastogenesis*

Bone-marrow-derived lymphocytes, after appropriate stimulation by antigen, RNA-immunogen, RNA or cell interactions undergo blastogenesis. In the mouse, these events may occur in the germinal centers of the lymphoid follicles. The proliferation of the immunocompetent progenitor cell and its migration to the red pulp where it differentiates into the plasma cell has been studied in detail (Hanna, 1965). The importance of the germinal center in the response is well documented (Cottier *et al.*, 1967). Dendritic reticular cells that constitute the stroma or matrix of active germinal centers have provided a definitive architectural configuration to localize antigen extracellularly within extensive cell-membrane infoldings (Szakal and Hanna, 1968). Such localization allows sequestering and maintenance of the antigen which relates directly to the proliferation of immunological competent progenitor cells. Thus, it is probable that the efficiency of the immune response and immunoglobulin biosynthesis *in vivo,* particularly blastogenesis, depends on an architectural site such as the germinal center for cell–cell and antigen–cell interactions.

B. Inhibition of the Immune Response with Ionizing Irradiation

Following the injection of an antigen, there is a short period of time during which the interactions described above occur, initiating the immune response. This period is referred to as the induction phase. Thus, we may consider the effects of ionizing radiation on the immune response in one of several ways: (1) exposure before the injection of antigen; (2) exposure during the induction phase; (3) exposure during the period of immunoglobulin biosynthesis (the production phase); (4) exposure during the secondary response. It has also been useful to study the effects of irradiation on cell populations in *in vitro* systems.

1. *Irradiation Exposure before Injection of Antigen*

When the irradiation is administered before the injection of an antigen, all indicators of the immune response are depressed. These parameters include decreased peak titers, prolonged latent period (that is, the time before specific antibody is detected in the blood), and a slow rise of the antibody population. Generally, irradiation is given shortly before antigen; however, when antigen is injected as late as 28 days after irradiation, the peak titer is still subnormal (Casarett, 1968). These observations indicate damage to the cell populations that interact with each other and with antigen and insufficient recovery before the antigen is given.

2. *Irradiation Exposure during the Induction Phase*

Irradiation during the induction phase results in a moderate to severe depression of the immune response depending upon the antigen and species of animal (Taliaferro *et al.*, 1964; Makinodan and Price, 1971). Damage during this period would be due to the effect on antigen-processing events and/or the subsequent cell–cell interactions.

3. *Irradiation Exposure during the Production Phase*

This phase of the immune response is generally radioresistant. During this period, the mature, nondividing, immunoglobulin-synthesizing cell (plasma cell) is the predominant type. This cell is highly radioresistant with no significant injury incurred at exposures of 10,000 R (Sado, 1969).

4. *Irradiation Exposure during the Secondary Response*

A secondary response results when a second dose of antigen is given at some time after the first. This response is characterized by a short latent period, a rapid rise in the appearance of specific antibody, and higher peak titers than is observed in the initial or primary response. Quantitatively the secondary response is more radioresistant than the primary response (Stoner and Hale, 1962). The secondary response depends upon the presence of committed immunocompetent cells that do not require antigen processing or cell–cell interactions (Šterzl, 1967) although blastogenesis and differentiation into plasma cells from memory cells occur.

5. *Irradiation of Cell Populations in Vitro*

The radiosensitivity of the cellular subpopulations that interact to develop the immune response has been determined by a variety of investigations. Although some generalizations can be made, the basic paradox is the fact that whereas exposure of the intact animal to doses of 800 R or much less, depending upon the species, will result in an inhibition of the immune response, doses of 4000 R and more are required to inactivate normal, previously unstimulated cells *in vitro*.

a. Macrophage. Phagocytosis and antigen degradation are relatively radioresistant. The relative engulfing capacity of peritoneal macrophages was not impaired after exposure to 50,000 R (Perkins *et al.*, 1966). These investigators suggest that the earliest effect of X-irradiation is the loss of the proliferative capacity of the macrophage but not the degradative or phagocytic activities.

Feldman and Gallily (1967) found that when macrophages were incubated *in vitro* with *Shigella* and injected into irradiated (550 R) mice, the animals produced circulating antibody, although antigen alone did not. Sensitized macrophages from normal animals elicited an immune response while macrophages from irradiated animals were unable to elicit a response. Similar results were obtained by Pribnow and Silverman (1967) using heat-treated bovine γ globulin as antigen. Earlier it was observed that bacteria and foreign erythrocytes were cleared from the circulation of X-irradiated animals at the normal rate, but the phagocytes did not degrade or retain these substances in a normal manner (Gordon *et al.*, 1955). All these investigators concluded that X-irradiation damages the capacity of macrophages to process antigen. On the other hand, Roseman (1969) reported that doses as high as 1000 R had no measurable effect on the immune function of adherent cells.

In addition to the impairment of the processing activity of the macrophage, X-irradiation can inhibit the production of macrophages from progenitor cells in the bone marrow and their recruitment at the site of a local lesion (Chandrasekhar *el al.*, 1971).

It is possible that antigen processing by macrophages may not be required for all antigens (Landy and Braun, 1969; Talmage *et al.*, 1970). However, the involvement of the macrophage in immunological phenomena including the retention and processing of antigen appears to be well established in many instances (Frei *et al.*, 1965; Franzl and McMaster, 1968a,b; Fishman and Adler, 1967; Landy and Braun, 1969; Bell and Dray, 1971). Another aspect in determining the role and radiosensitivity of the macrophage is a consideration of their location in the body and the associated architecture in that location. Thus, macrophages located in lymphoid follicles were found to be different from those in other areas (Hunter and Wissler, 1967) while morphological and physiological characteristics distinguish macrophages in various other organs and tissues (Pearsall and Weiser, 1970).

b. Bone-Marrow-Derived Nonadherent Cell. The radiosensitivity of this cell is probably the most critical in the inhibition of the immune response since these cells will proliferate and differentiate to plasma cells, or immunoglobulin-synthesizing cells (Mitchell and Miller, 1968) and at a rapid rate (Rowley *et al.*, 1968). The effect of X-irradiation on these cells may be considered in either of two ways: (a) cessation of proliferation of the progenitor cells resulting in a subnormal population that can be stimulated and (b) the impairment of blastogenesis and proliferation after stimulation resulting in a subnormal population that can differentiate into plasma cells. Since the most drastic inhibition of the immune response occurs when X-irradiation is given before the antigen, the population of the bone-marrow-derived cells available for antigen stimulation or cell–cell interactions must be quite small and/or impaired in their capacity to be stimulated.

c. Thymus-Derived Nonadherent Cells. Thymocytes can restore the *in vitro* immune response of spleen cells that have been thymectomized, heavily irradiated, and then given syngeneic bone-marrow cells (Mosier *et al.*, 1969). The addition of thymocytes to cultures of unseparated, separated, or recombined spleen cell populations (adherent and nonadherent) from irradiated mice failed to restore the response. Although these results indicate that the X-ray sensitive population was the bone-marrow-derived cell, the results suggest that the interaction of thymus-derived and bone-marrow-derived cells may be impaired.

d. Bursa-Equivalent Influence. Information regarding the cell type or other factors contributed by the bursa-equivalent is limited, although this organ is essential for immunoglobulin synthesis. The shielding of the appendix from X-irradiation was found to allow complete induction of the immune response in rabbits (Taliaferro *et al.*, 1964). Thus the influence of the bursa-equivalent may be considered as radiosensitive.

C. Enhancement of the Immune Response with Ionizing Irradiation

Under certain conditions, exposure to ionizing radiation will provoke an enhanced immune response. Generally, such enhancement occurs when the antigen is administered shortly before or after exposure to a sublethal dose of radiation to the whole body, or to a supralethal dose locally. The phenomenon was first observed by Manoukhine (1913) and confirmed repeatedly by others (Makinodan, 1966; Simić *et al.*, 1965; Taliaferro *et al.*, 1964; Simić and Petrović, 1971).

The maximum immune response that an animal undergoes reflects an immunological expression that is only a fraction of its full potential (Santos and Owens, 1966). The extent of this expression would depend upon the availability of tissue space that would limit the proliferation of stimulated cells, that is, the number of plasma cells synthesizing immunoglobulins would be limited by the space in the red pulp. In the irradiated animal, the cell destruction in the spleen would allow an expansion of space in which surviving cells proliferate (Simić and Petrović, 1971). In addition, cell destruction would provide an enriched environment for proliferation (Makinodan and Price, 1971).

D. Restoration of the Immune Response after Irradiation

Immunoglobulin-producing capacity of an irradiated animal can be restored by one of several materials. In general, nucleic acids will restore the immune response although colchicine, kinetin, and 3-indoleacetic acid are effective (see Abramoff and LaVia, 1970). The restorative activity of nucleic acids depends upon the state of the material in that polymerized DNA or RNA, mixtures of nucleosides, nucleotides, and nucleotide diphosphates are ineffective, whereas digests containing reasonably high-molecular-weight oligonucleotides may restore the response completely.

It has been suggested that nucleic acids provide an enriched medium for cells that survive irradiation to affect survival, differentiation, and mitotic activity (Simić *et al.,* 1965; Makinodan and Price, 1971). Restoration is probably related to the repair of radiation-damaged genetic apparatus of the cell.

Colchicine is the most active of all the restorative agents. It is believed to be cytotoxic *in vivo,* resulting in cell death with the release of partially degraded nucleic acid.

E. Theoretical Considerations

On the basis of the preceding discussions, we have described the relative radiosensitivity of cell types involved in the immune response in Fig. 8. Several pertinent references are given in the figure but are not presumed to be complete in any way. The evidence indicates that the bone-marrow-derived lymphocyte (nonadherent cell) is the most susceptible to irradiation damage. Impairment of the genetic apparatus, however, must occur equally as often in all cell types on a mere hit-probability basis. Other possibilities must then be considered.

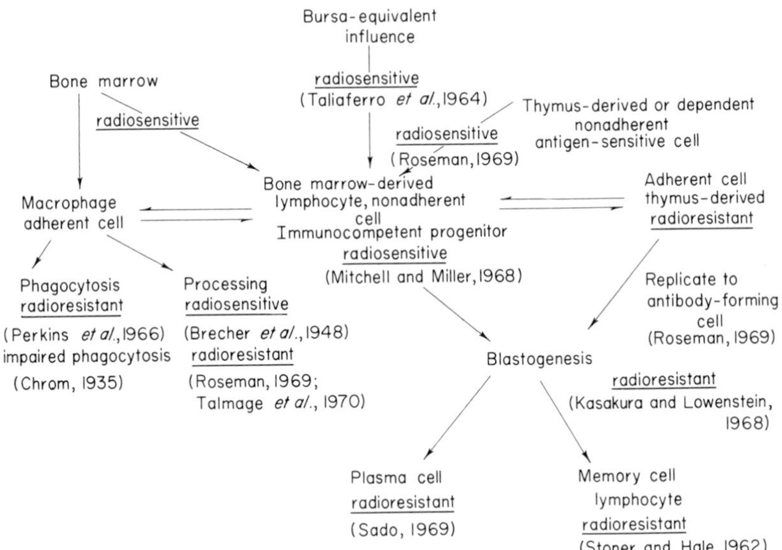

Fig. 8. Relative radiosensitivity of cells or cell–cell interactions in the immune response.

Radiosensitivity is related primarily to the proliferative activity of progenitor cells. As proposed in earlier sections of this chapter and from discussions in Chapter 4, the most sensitive period in the cell cycle is the G_2 period. During this phase of the cell cycle, specific synthesis of the components necessary for mitosis takes place. It is also during this phase that repair mechanisms are most operative with regard to the excision of damaged DNA and polynucleotide ligase activity.

We suggest that the inhibition of the immune response is due to damage to the macromolecules, particularly DNA, of cells that are proliferating (stem cells, progenitor cells, antigen-stimulated cells). The generation time of these cells is so rapid that repair mechanisms are either nonoperable or cannot keep pace with the events of mitosis. The damage to cells in G_2 must involve impairment in the synthesis of the mitotic apparatus (proteins, spindle, etc.). Thus mitosis would not occur, or if it did occur before adequate repair, the progeny would die. This might be called unbalanced intracellular homeostasis, resulting in cell death.

The restorative effects of colchicine are probably directly related to restored intracellular homeostasis. Colchicine is not only a cytotoxic agent but a substance that causes mitotic arrest, or retardation. If mitosis were delayed, repair mechanisms would be accomplished before cell division. Repair includes not only DNA but membranes, sulfhydryl-containing proteins, and RNA necessary for regulation.

It is quite possible that the restorative effects of nucleic acids, or cellular extracts containing nucleic acids may be to provide polynucleotide sequences for insertion into damaged areas of DNA by ligase activity. A suggestion for this repair and/or synthesis following irradiation was discussed in Chapter 4 and in the investigations of Shepard (1965). He found abnormally high amounts of DNA (3 times normal) in *Tetrahymena* prior to the first division after irradiation.

The committed cell in the immune response proliferates and differentiates into the plasma cell which is relatively radioresistant. As much as 60% of the protein secreted by this cell may be accounted for as specific antibody molecules (Helmrich *et al.*, 1961). The synthesizing activity of this cell must require large amounts of specific mRNA. During differentiation, unique redundant genes coding for the specific immunoglobulin polypeptide chains may have tandemly replicated to provide the genetic information necessary for large amounts of the corresponding mRNA. Using RNA–DNA hybridization techniques, Little and Donahue (1970) showed increased association between DNA and RNA from lymph nodes of animals immunized to 2,4-dinitrophenyl bovine γ globulin (2,4-DNP-BGG) or 2,4-DNP egg albumin as compared to hybridization with DNA and/or RNA from lymph nodes of nonimmunized animals, or from liver. They concluded that lymph nodes from immunized animals contained unique sequences of DNA that were redundant. If tandem replicates are present, it seems likely that a much higher incidence of hits would be necessary to inactivate all transcription for specific antibody synthesis.

Another, although more unlikely, possibility of tandem replicates might be considered here. Callan (1967) postulated that in the genomes of chromosomal organisms (eukaryotes) each unit of genetic information is serially repeated. In this hypothesis, the terminal unit serves as a master sequence within which recombinational events can occur and which may not at any time engage directly in RNA synthesis. The "master" is followed by "slave" sequences that only reflect recombination in the master gene but can be transcribed to RNA message. This hypothesis was extended by Whitehouse (1967), who suggested that the chromosome has the form of a cycloid, the loop of which corresponds to a set of "slaves." In this modification, the master is the only gene integrated into the chromatid and susceptible to recombinational events.

This hypothesis may be extended to provide an explanation for differentiation and the relative radioresistance of the differentiated cell. A master and slave assortment of genes gives extensive genetic information of exacting specificity. Each of these many copies would have to be inactivated with irradiation in order for the cell to be inoperative. Thus the differentiated cell is much more radioresistant when compared with the progenitor, undifferentiated cell possessing only the master.

The notion that tandem repetitions of some kind exist in eukaryotic DNA is supported by Thomas *et al.* (1970), who were able to induce circle formation with fragmented eukaryotic DNA. Whether these repetitions are exact, as would be required by the models of Callan and Whitehouse, has not been demonstrated however. The data of Britten and Kohne (1968) on thermal stability and hyperchromicity of reassociated fragments indicate that repetitions are inexact. It may be, however, that several different sets of exact tandem duplications are sufficiently closely related to reassociate with one another.

VII. Radiation and Cancer

A. Introduction

It has been known for many years that exposure to large doses of ionizing radiation promotes and increases the incidence of cancer. Some of the most common cancers which have been associated with radiation exposure are those of the skin, lungs, bone, liver, spleen, and blood (leukemia) (reviewed by Upton, 1968).

The mechanism of the induction of cancer by ionizing radiation is not known. This is not surprising in view of the fact that the mechanisms of induction of cancer by much more specific agents (oncogenic viruses, for example) are also obscure. Clarification of the action of these apparently simpler agents is likely to precede an understanding of the mechanism of radiation carcinogenesis. Thus the study of radiation induction of cancer must be more concerned with the general features of cancer tissues, and the ways in which they differ from normal tissues, than with radiation per se. Once these differences are better characterized, it will become feasible to inquire into how ionizing radiation might account for their induction.

The general problem is complicated by the fact that "cancer" is a term used to denote a wide variety of pathological conditions, and it is difficult to make any generalizations that will be true of all cancers. Numerous morphological and biochemical changes do take place upon induction of the cancerous state.

Among the changes that occur in malignant transformation are the disappearance of contact inhibition and appearance of surface antigens on the tumor cells, which were either absent or masked in the normal cell. In the case of virus-induced tumors, all tumors induced by the same virus have neo-antigens in common, suggesting close similarity of the oncogenic process in each case. In tumors induced by the nonspecific oncogenic agents (chemical carcinogins and radiation), a series of tumors induced by the same agent display no similarity in the neo-antigens present. These results suggest that the mode of action of radiation and chemical carcinogens may be more complex than that of the viruses. It will be of value to inquire into the general nature of the biochemical changes that occur in cancer cells and then to consider possible mechanisms of carcinogenesis. Finally we shall consider the role of radiation in cancer.

B. Biochemical Characteristics of Cancer Cells

The characteristics of cancer cells described in the preceding paragraph indicate that neoplastic transformation involves, among other things, changes in the properties of the cell surface. These characteristics do not, however, in themselves, give us much insight into the molecular basis of the transformation. For this purpose we must inquire in more detail into the biochemical properties of cancer.

1. *DNA*

The DNA complement of cancer cells may differ from that of normal cells in several ways. Perhaps the most dramatic effect is an alteration of chromosome number. While altered chromosome number is a common characteristic of cancer cells, its significance and origin are not clear (Kirby, 1967). Detailed investigations of the nucleotide sequence distributions in the DNA's of rat liver and rat hepatoma have been reported by Britten and Rake (1969). They studied the reassociation behavior of both DNA's in order to determine whether certain DNA sequences were amplified in the hepatoma (in this case a rapidly proliferating tumor with a chromosome number of 73 as opposed to the normal value of 42). They were able to detect no differences between the two tissues. While these results strongly suggest that the overall sequence composition of the DNA of normal and tumor tissues is similar, the experimental techniques are open to two reservations:

1. The amplification of one or a few genes would not have been detected.
2. The experiment was confined to an examination of the relative abundance of various frequency classes. More elaborate experiments would be required to determine whether the sequences represented in the various frequency classes are the same or different.

The impulse to look so critically at the results reported by Britten and Rake would not be so strong were it not for several recent reports indicating differences in the DNA sequence distributions in different tissues or cell types in an organism. The first of these differences was the observation of Brown and David (1969) that cistrons coding for ribosomal RNA are amplified in oocytes of the African clawed toad *Xenopus laevis*. This is, in a sense, a special case, in that the amplified DNA is not a part of a chromosome.

The second observation, which we wish to point out in this regard, is the difference in DNA content of spleen and liver DNA of the rabbit discovered by Little and Donahue (1970). They found that the spleen contains more sites homologous to RNA synthesized by lymph nodes of immunized rabbits than does the liver. Similar experiments, using a tissue-specific RNA as reagent, might reveal DNA differences between normal and tumor tissues, not detected by the techniques of Britten and Rake (1969). Finally, Krueger and McCarthy (1970) have found that various BalbC mouse myelomas contain repeated copies of certain nucleotide sequences and that the sequences repeated are different in the different myelomas. Read in the light of the results of McCarthy and Hoyer (1964), these results indicate that, while the same nucleotide sequences are found in the repetitious class in most tissues, the frequency at which different classes are repeated may vary considerably. Obviously this is a fertile field for future investigation.

Another, and more tractable, case of DNA alteration in cancer cells occurs when a normal cell is transformed by a DNA-containing oncogenic virus. This subject has recently been reviewed by Green (1970). It has been shown repeatedly that the DNA of such a virus may be incorporated into the host genome and replicated along with the host's DNA. The host's transcriptive apparatus does not appear to distinguish between indigenous and foreign DNA, since RNA homologous to the viral DNA is found in transformed cells.

2. *Protein*

It is perhaps obvious to point out there are differences in the protein distributions of normal and cancer cells. In some instances these differences have been rather extensively characterized; see, for example, the studies on the ribonucleases, reviewed by Roth (1967). Of perhaps more interest in considering the origin of neoplasia are differences in nuclear proteins (Busch and Mauritzen, 1967). Nuclear proteins, both histone and nonhistone proteins, have been implicated as important factors in the control of transcription in mammalian cells. While the relative importance of nuclear protein and other chromosomal elements in the transcriptive process remains to be clarified, it would appear that this, too, represents a promising area of investigation.

3. *RNA*

Intermediate between DNA and protein in the information transfer system is RNA. The study of RNA in cancer cells has taken place along several, more-or-less unrelated pathways.

A great deal of work has been done on the transcription of RNA from the viral genome incorporated into cells transformed by oncogenic viruses (reviewed by Green, 1970). In terms of volume, considerably less information is available concerning the transcription (and its control) of host-specific RNA in normal and transformed cells. The literature on this subject is surprisingly small. Studies by Drews *et al.* (1968) revealed that differences do exist between the RNA's from the two sources. This observation is consistent with the observations by others that there are RNA differences between tissues and between regenerating and nonregenerating forms of the same tissue (McCarthy and Hoyer, 1964; Church and McCarthy, 1967a,b). Moreover, Chiarugi (1969) has obtained qualitatively similar results, using a more discriminating technique, with rat livers and hepatomas. Species of nuclear RNA that do not appear in the cytoplasms have also been observed by Shearer and McCarthy (1967) and by Soeiro and Darnell (1969). Experiments reported by Roche *et al.* (1969) revealed differences in the nuclear RNA of several pairs of normal–neoplastic tissues derived from human sources. Their measurements reflected principally differences in size, however, and are not necessarily related to sequence differences.

C. Theories of Carcinogenesis

Oncogenesis could conceivably occur at any of the three levels of macromolecular synthesis: DNA replication, transcription into RNA, and translation into protein. Most attention has been focused upon the first two of these alternatives, and this attention has given rise to two schools of thought.

The first holds that oncogenesis is the result of a somatic mutation. According to this view, the oncogenic agent induces a mutation in a normal cell, leading to a defect in some part of the cell's synthetic activity. The result of this defect, then, is both expressed in the mutant cell and transferred to each progeny cell at mitosis.

The second point of view holds that oncogenesis is the result of a disturbance not of the primary genetic material, but in the control of its expression. Among supporters of this hypothesis, there is no general agreement as to just what genes are turned on or off during oncogenesis or how the control switch occurs. The school has numerous subschools. One of the boldest suggestions made by the advocates of a control hypothesis is that of Huebner and Todaro (1969), who believe that most if not all human cancers are the result of the activation of a gene which codes for a particular type of RNA virus. Activation of this gene (the oncogene) results in both production of the virus and changes in cellular control characteristic of the cancerous state.

For some time, the somatic mutation hypothesis held the center of the stage in cancer research, and it is not yet to be dismissed. More recently, however, several lines of evidence have moved opinion in the direction of the control hypothesis.

1. *Somatic Mutation Hypothesis*

One of the early reasons for advancing the somatic mutation theory was "the firmly established fact that the malignant transformation is permanent" (Alexander, 1957). But this assertion has proved too general, and reversion of virus-transformed cells upon loss of the viral genome has been observed (Marin and Littlefield, 1968; Marin and MacPherson, 1969). Moreover, the expression of particular viral genes is known to be required for the establishment and maintenance of the transformed state induced by both RNA-containing (G. S. Martin, 1970) and DNA-containing oncogenic viruses (Dulbecco and Eckhart, 1970).

A more spectacular demonstration of probable lack of somatic mutation in a cancer cell was provided when it was shown that the nucleus of a frog renal adenocarcinoma transplanted into a normal enucleate egg gave rise to an advanced embryo (King and McKinnell, 1960). Two points should be made about this result. First, since the nucleus directed the formation of normally differentiated tissues, its genome cannot be considered to be a "cancer genome" containing a specific cancer-inducing somatic mutation. Second, the fact that the embryo reached an advanced stage suggests that the nucleus had a developmental potential more like that of an embryonic cell than that of a more highly differentiated cell. This result may be directly related to the finding in several laboratories of antigenic determinants common to fetal and tumor tissues (Gold and Freedman, 1965; Stonehill and Bendich, 1970; Coggin *et al.*, 1970a, 1971).

It can also be argued that agents that cause mutations also cause cancer. But such a correlation as this must be examined more critically. Mutations, even in higher organisms, are usually expressed soon after induction (Drake, 1969). Similarly, many cancers develop soon after the introduction of the putative "mutagenic" agent. Other cancers, however, develop only many years after exposure of the subject to the inducing agent. These results could be compatible with the somatic mutation theory if either of the following hypotheses were true:

1. The mutated cell is rendered incapable of proliferation for some time after the mutagenic event, and requires a change in the physiological or immunological state of the animal (e.g., aging).

2. The mutated cell divides normally, but does not express its neoplastic nature until some change occurs in the physiological state of the animal. Missense mutants that might yield gene products differing only slightly in structure from the wild-type gene product might play such a role.

Either of these possibilities is, of course, quite conceivable.

2. *Control Hypothesis*

It was stated earlier that one can construct many models of the way in which a breakdown of normal cellular regulatory functions might lead to malignant transformation. In this section we shall stress the role of RNA in such a process.

It is characteristic of embryonic development that stimuli received by a cell can cause it, at some later time, to become differentiated in a certain way. These stimuli can be very subtle; the earth's gravitational field, for example, can provide an environmental asymmetry that determines the developmental pattern of an organism. We find it useful to look upon the oncogenic process in much the same way. We suppose that the oncogenic agent affects the cell in a manner analogous to that environmental asymmetry. Once this occurs, it sets in motion a sequence of events that, eventually, leads to the production of a recognizable cancer cell. The process may be long or short depending on the system involved. This point of view is noncommittal as to the particular events that follow one another, but we shall consider some more specific hypotheses, all of which contain the basic elements outlined above.

The oncogene hypothesis is a case in point. Here it is postulated (Huebner and Todaro, 1969) that the information necessary to produce a neoplastic conversion directly is contained in the chromosomal DNA of all cells, and that it is vertically transmitted from parent to progeny. It is supposed, however, that this genetic information is normally repressed. The repression may be removed spontaneously, or it may be induced by changes in the cellular environment, such as aging or the introduction of "carcinogenic" agents. Once this information is derepressed, a malignant transformation follows.

A second model can be constructed using the general theory for eukaryotic regulation proposed by Britten and Davidson (1969), which has already been mentioned in connection with antibody synthesis. The multitiered sets of controls envisioned in this model allow great flexibility. One could imagine, for example, that a carcinogenic agent might, directly or indirectly, trigger a sensor gene into initiating certain activator RNAs. These, in turn, would cause the production of RNA and/or protein under the direction of one or more producer genes. These products, in turn, could act upon other sensor genes, and initiate the whole process over again. Thus, a single molecular event could trigger a cascade of reactions culminating in the diversity of RNAs found in cancer cells.

Before leaving the general area of control processes, it should be pointed out that controls at levels other than the transcriptional may be involved. It has been quite firmly established that eukaryotic ribosomal RNA is subject to elaborate posttranscriptional processing (Darnell *et al.*, 1970; Grierson *et al.*, 1970). It appears, too, that the heterogeneous nuclear RNA undergoes posttranscriptional processing (Scherrer *et al.*, 1970) to be converted into chromosomal RNA (Mayfield *et al.*, 1971), which appears to have a role in transcriptional specificity (Bekhor *et al.*, 1969). Whether posttranscriptional controls have a role in malignant transformation remains to be seen.

D. Radiation and Cancer

What do we know about radiation and cancer, and what can radiation research contribute to solving the cancer problem? Radiation, as it is related to cancer, has been called a "double-edged sword" (Alexander, 1957), in that it can be either a "cause" or a "cure." We shall not concern ourselves here with the therapeutic use of radiation in cancer, but rather with its role in the oncogenic process.

The use of radiation to produce mutations for genetic studies is as old as artificial mutagenesis itself (Muller, 1927). Thus it would be comfortable to assume that radiation induces cancer through a mutagenic mechanism. Such a view is not, however, universally accepted. Leukemia, for example, is readily induced by chronic doses of ionizing radiation; at the same time, leukemias are among the cancers most closely linked to a viral etiology. Some investigators believe that viruses may be the initiating agents for radiation-induced leukemia (for review, see Casarett, 1968). It is argued that the radiation promotes the expression of exogenous (Casarett, 1968) or endogenous virus (Huebner and Todaro, 1969). The way in which the radiation may act to promote viral transformation may be through a depression of the host's immunity to exogenous virus (Casarett, 1968) or, as mentioned above, through induction of the oncogene or through some as yet unknown mechanism. Other evidence implicating ionizing radiation as a promoter of viral carcinogenesis includes the finding that preexposure of neonatal hamsters to X-rays increases the incidence of SV40-induced tumors (Coggin *et al.*, 1970b). Similar results have been obtained with cells in culture (Stoker, 1963; Kouri and Coggin, 1968; Coggin, 1969).

In spite of some modest success in tracking down the relationship between radiation and cancer, no very firm generalizations can be drawn. In particular, radiation carcinogenesis does not appear to be on the verge of a major contribution to our knowledge of the molecular basis of cancer. Watson (1965) has argued on purely technical grounds that it is not useful to try to characterize the early events in oncogenesis by studying cancers induced by chemicals or by ionizing radiation. This objection, taken along with the observations on nonspecificity made at the beginning of this section, lead us to conclude that an elucidation of the mechanism of radiation carcinogenesis will probably follow, rather than precede, an understanding of the general nature of malignant transformation.

E. Xeroderma Pigmentosum

In spite of our meager knowledge concerning most radiation-induced (promoted) cancers, recent discoveries concerning a rare skin cancer in man, xeroderma pigmentosum, are beginning to establish a macromolecular basis for this radiation-promoted cancer. This section deals with these new discoveries.

Xeroderma pigmentosum (XP) is a rare autosomal recessive disease. The skin of homozygous affected individuals is extremely sensitive to intense sunlight and ultraviolet light (Seguira *et al.*, 1911; McKusick, 1966). The disease is progressive, starting in children at about age 3 (a time when children begin to venture out-of-doors). It begins with the formation of freckles, which lead to atrophy and a number of keratoses of the skin (Rook *et al.*, 1968). These progressive changes lead to the eventual ulceration of the epidermis and the formation of basal cell and squamous cell metastatic carcinomas. Death of affected individuals occurs before the age of 30.

Tissue culture cell lines have recently been established from patients with XP, and several different biochemical experiments have been performed in an attempt to elucidate the mechanism responsible for this ultraviolet light (UV)-promoted skin cancer (Cleaver, 1966, 1967, 1969; Setlow *et al.*, 1969). These studies have involved comparisons of the processes of repair replication and the production and fate of UV-induced pyrimidine dimers in normal human cells and in XP-affected cells. In order for us to describe these studies further, we must first review the normal repair processes that are known to occur in UV-irradiated mammalian cells.

The induction of thymine dimers in mammalian cells was first reported by Trosko *et al.* (1965) in studies of Chinese hamster cells *in vitro*. It was observed in this study and in subsequent reports (Klimek, 1965, 1966; Trosko and Kasschau, 1967) that UV-induced pyrimidine dimers were not subject to the photoreactivation and excision processes that had been demonstrated to occur in bacterial cells. Subsequent studies confirmed that cells from placental mammals are indeed not capable of photoreactivation of pyrimidine dimers (Cleaver, 1966; Cook and McGrath, 1967; Cook, 1970), although photoreactivation does occur in certain marsupial cell lines (Cook and Regan, 1969). Cook (1970) concluded that photoreactivating enzyme probably does not exist in placental mammals. On the other hand, more detailed investigations have now shown that normal human cells *in vitro* are capable of excision processes (Regan *et al.*, 1968; Chapter IV, Fig. 4 this volume) and that can perform normal repair replication (Cleaver, 1968).

A key finding concerning XP cells in culture was the fact that these cells were found to be defective in their ability to perform normal repair replication (Cleaver, 1968). More recent evidence from XP fibroblasts in tissue culture indicates that these cells are only defective in the initial stage of DNA repair. (Cleaver, 1969; Chapter 1, Fig. 16 and Chapter 4, Fig. 4B this volume; Setlow *et al.*, 1969). It was concluded by Cleaver that other repair processes (other enzymatic steps) were not defective in these cells if the initial scission of the polynucleotide chain was mediated in some other way, such as irradiation of DNA in which thymine has been replaced by bromouracil (BUdR). Studies by Setlow and co-workers suggest "that XP cells are deficient in a functional form of an ultraviolet-specific endonuclease essential to the initiation of repair." They were cautious in their interpretation because they are uncertain whether the endonuclease may be present and propose the possibility that other mechanisms might be interfering "with the proper functioning of an otherwise normal enzyme." Both Cleaver and Setlow speculate whether failure of XP cells to excise UV-damaged regions in their DNA accounts for neoplastic transformation of these cells into cancer cells. Cleaver (1969) was quick to point out that other malignant cells, particularly HeLa cells, are capable of normal UV repair replication and that this "shows that defective repair is not essential for carcinogenesis." In view of the diversity of processes that are referred to as carcinogenesis, however, such a comparison is not particularly convincing.

Acknowledgments

Preparation of this chapter was supported in part by a grant from the American Cancer Society to W. S. R., by N. I. H. Biomedical Science support grants to the University of Tennessee (C. J. W. and G. L. W.) and by an American Cancer Society Institutional Grant to the University of Tennessee (C. J. W. and W. S. R.). We are grateful to Dr. Arthur Brown for critical reading of parts of the manuscript, to Miss Sarah Fritts for help in preparation, and to Miss Linda Parker for typing the successive versions.

References

Abdou, N. I., and Richter, M. (1970). *Advan. Immunol.* **12**, 201.

Abramoff, P., and LaVia, M. F. (1970). "Biology of the Immune Response." McGraw-Hill, New York.

Alexander, P. (1957). "Atomic Radiations and Life." Penguin Books, London.

Allen, R. G., Brown, F. A., Logie, L. C., Rovner, D. R., Wilson, S. G., and Zellmer, R. W. (1960). *Radiat. Res.* **13,** 532.

Bekhor, I., Kung, G., and Bonner, J. (1969). *J. Mol. Biol.* **39,** 351.

Bell, C., and Dray, S. (1969). *J. Immunol.* **103,** 1196.

Bell, C., and Dray, S. (1971). *Science* **171,** 199–201.

Bender, M. A. (1969). *Advan. Radiat. Biol.* **3,** 215.

Blondel, B., and Tolmach, L. J. (1965). *Exp. Cell Res.* **37,** 497.

Bollum, F. J., Auderegg, J. W., McElya, A. B., and Potter, V. R. (1960). *Cancer Res.* **20,** 138.

Bond, V. P., Fliedner, T. M., and Archambeau, J. O. (1965). "Mammalian Radiation Lethality." Academic Press, New York.

Brent, T. P., Butler, J. A. V., and Cruthorn, A. R. (1965). *Nature* (*London*) **207,** 176.

Brewen, J. G. (1965). *Int. J. Radiat. Biol.* **9,** 391.

Britten, R. J., and Davidson, E. H. (1969). *Science* **165,** 349.

Britten, R. J., and Kohne, D. E. (1968). *Science* **161,** 529.

Britten, R. J., and Rake, A. V. (1969). *Carnegie Inst. Wash., Yearb.* **67,** 325.

Brown, D. D., and David, I. (1969). *Annu. Rev. Genet.* **3,** 127.

Busch, H., and Mauritzen, C. M. (1967). *Methods Cancer Res.* **3,** 392.

Callan, H. G. (1967). *J. Cell Sci.* **2,** 1.

Carlson, J. G. (1967). *Radiat. Res.* **31,** 573.

Carlson, J. G. (1969). *Radiat. Res.* **37,** 15.

Casarett, A. P. (1968). "Radiation Biology," Prentice-Hall, Englewood Cliffs, New Jersey.

Cattaneo, S. M., Quastler, H., and Sherman, F. G. (1961). *Nature* (*London*) **190,** 923.

Cebra, J. J. (1969). *Bacteriol. Rev.* **33,** 159.

Chan, E. L., Mishell, R. I., and Mitchell, G. F. (1970). *Science* **170,** 1215.

Chandrasekhar, S., Shima, K., Dannenberg, A. M., Kambara, T., Fabrikant, J. I., and Roessler, W. G. (1971). *Infec. Immunity* **3,** 254.

Chiarugi, V. P. (1969). *Biochim Biophys. Acta* **179,** 129.

Chu, E. H. Y., Giles, N. H., and Passano, K. (1961). *Proc. Nat. Acad. Sci. U. S.* **47,** 830.

Church, R. B., and McCarthy, B. J. (1967a). *J. Mol. Biol.* **23,** 459.

Church, R. B., and McCarthy, B. J., (1967b). *J. Mol. Biol.* **23,** 477.

Cleaver, J. E. (1966). *Nature* (*London*) **209,** 1317.

Cleaver, J. E. (1967). *Radiat. Res.* **30,** 795.

Cleaver, J. E. (1968). *Nature* (*London*) **218,** 652.

Cleaver, J. E. (1969). *Proc. Nat. Acad. Sci. U. S.* **63,** 428.

Coggin, J. H., Jr. (1969). *J. Virol.* **3,** 458.

Coggin, J. H., Jr., Ambrose, K. R., and Anderson, N. G. (1970a). *J. Immunol.* **105,** 524.

Coggin, J. H., Jr., Harwood, S. E., and Anderson, N. G. (1970b). *Proc. Soc. Exp. Biol. Med.* **134,** 1109.

Coggin, J. H., Jr., Ambrose, K. R., Bellomy, B. B., and Anderson, N. G. (1971). *J. Immunol.* **107,** 526.

Congdon, C. C. (1959). *Progr. Hematol.* **2,** 21.

Congdon, C. C. (1971). *Science* **171,** 1116.

Cook, J. S. (1970). *In* "Photophysiology" (A. C. Giese, ed.), Vol. 5, p. 191. Academic Press, New York.

Cook, J. S., and McGrath, J. R. (1967). *Proc. Nat. Acad. Sci. U. S.* **58,** 1359.

Cook, J. S., and Regan, J. D. (1969). *Proc. Nat. Acad. Sci. U. S.* **58,** 2274.

Cooper, M. D., Peterson, R. D. A., and Good, R. A. (1966). *In* "Phylogeny of Immunity" (R. T. Smith, P. A. Miescher, and R. A. Good, eds.), p. 243. Univ. of Florida Press, Gainesville.

Cottier, H., Odartchenko, N., Schindler, R., and Congdon, C. C., eds. (1967). "Germinal Centers in Immune Responses." Springer Publ., New York.

Crippa, M. (1966). *Exp. Cell Res.* **42,** 371.

Darnell, J. E., Pagoulatos, G. N., Lindberg, U., and Balint, R. (1970). *Cold Spring Harbor Symp. Quant. Biol.* **35,** 555.

Deschner, E. E., and Gray, L. H. (1959). *Radiat. Res.* **11,** 115.

Detrick, L. E., Latta, H., Upham, H. C., and McCandless, R. (1963). *Radiat. Res.* **19,** 447.

Dewey, W. C., and Humphrey, R. M. (1962). *Radiat. Res.* **16,** 503.

Dewey, W. C., and Humphrey, R. M. (1964). *Exp. Cell Res.* **35,** 262.

Dewey, W. C., Humphrey, R. M., and Sedita, B. A. (1966). *Biophys. J.* **6,** 247.

Doida, U., and Okada, S. (1967). *Nature (London)* **216,** 272.

Doida, U., and Okada, S. (1969). *Radiat. Res.* **38,** 237.

Drake, J. (1969). *Annu. Rev. Genet.* **3,** 247.

Drews, J., Brawerman, G., and Morris, H. P. (1968). *Eur. J. Biochem.* **3,** 284.

Dulbecco, R., and Eckhart, W. (1970). *Proc. Nat. Acad. Sci. U. S.* **67,** 1775.

Elkind, M. E., and Whitmore, G. F. (1967). "The Radiobiology of Cultured Mammalian Cells." Gordon & Breach, New York.

Evans, H. J. (1962). *Int. Rev. Cytol.* **12,** 221.

Feldman, M., and Gallily, R. (1967). *Cold Spring Harbor Symp. Quant. Biol.* **32,** 415.

Fishman, M., and Adler, F. L. (1967). *Cold Spring Harbor Symp. Quant. Biol.* **32,** 343.

Franzl, R. E., and McMaster, P. D. (1968a). *J. Exp. Med.* **127,** 1087.

Franzl, R. E., and McMaster, P. D. (1968b). *J. Exp. Med.* **127,** 1109.

Frei, P. C., Benacerraf, B., and Thorbecke, G. J. (1965). *Proc. Nat. Acad. Sci. U. S.* **53,** 20.

Froese, G. (1966). *Int. J. Radiat. Biol.* **10,** 353.

Galavazi, G., and Bootsma, D. (1966). *Exp. Cell Res.* **41,** 438.

Gally, J. A., and Edelman, G. M. (1970). *Nature (London)* **227,** 341.

Gelfant, S. (1962). *Exp. Cell Res.* **26,** 395.

Gleich, G. J., Bieger, R. C., and Stankievic, R. (1969). *Science* **165,** 606.

Glick, B. (1970). *BioScience* **20,** 602.

Gold, P., and Freedman, S. (1965). *J. Exp. Med.* **122,** 467.

Goldstein, A. L., Slater, F. D., and White, A. (1966). *Proc. Nat. Acad. Sci. U. S.* **56,** 1010.

Goldstein, A. L., Banerjee, S., and White, A. (1967). *Proc. Nat. Acad. Sci. U. S.* **57,** 821.

Gordon, L. E., Cooper, D. B., and Miller, C. P. (1955). *Proc. Soc. Exp. Biol. Med.* **89,** 577.

Gottlieb, A. A., Glišin, V. R., and Doty, P. (1967). *Proc. Nat. Acad. Sci. U. S.* **57,** 1849.

Green, M. (1970). *Annu. Rev. Biochem.* **39,** 701.

Grierson, D., Rogers, M. E., Sartirana, M. L., and Loening, U. E. (1970). *Cold Spring Harbor Symp. Quant. Biol.* **35,** 589.

Halpern, M. S., and Koshland, M. E. (1970). *Nature* **228,** 1276.

Handford, S. W. (1960). *Radiat. Res.* **13,** 712.

Handford, S. W., Johnson, P. W., Scholtes, R. J., and Duny, M. S. (1961). *Radiat. Res.* **15,** 734.

Hanna, M. G. (1965). *Int. Arch. Allergy Appl. Immunol.* **26,** 230.

Helmrich, E., Kern, M., and Eisen, H. N. (1961). *J. Biol. Chem.* **236,** 464.

Howard, A., and Pelc, S. R. (1951). *Exp. Cell Res.* **2,** 178.

Howard, A., and Pelc, S. R. (1953). *Heredity, Suppl.* **6,** 261.

Hsu, T. C., Dewey, W. C., and Humphrey, R. M. (1962). *Exp. Cell Res.* **27,** 441.

Huebner, R. J., and Todaro, G. J. (1969). *Proc. Nat. Acad. Sci. U. S.* **64,** 1087.

Hunter, R. L., and Wissler, R. W. (1967). *J. Reticuloendothel. Soc.* **4,** 444.

Ishizaka, K., and Ishizaka, T. (1968). *J. Allergy* **42,** 330.

Jerne, N. K. (1967). *Cold Spring Harbor Symp. Quant. Biol.* **32,** 591.

Johnson, T. C., and Holland, J. J. (1965). *J. Cell Biol.* **27,** 565.

Jureziz, R. E., Thor, D. E., and Dray, S. (1970). *J. Immunol.* **105,** 1313.

Kabat, E. A. (1968). "Structural Concepts in Immunology and Immunochemistry." Holt, Rinehart and Winston, New York.

Kasakura, S., and Lowenstein, L. (1968). *J. Immunol.* **101,** 12.

Kasten, F. H., and Strasser, F. F. (1966). *Nature* (*London*) **211,** 135.

Kim, J. H., and Evans, T. C. (1964). *Radiat. Res.* **31,** 129.

King, T. J., and McKinnell, R. G. (1960). *In* "Cell Physiology of Neoplasia" (14th Annual Symposium on Fundamental Cancer Research, M. D. Anderson Hospital and Tumor Institute), p. 591. University of Texas Press, Austin.

Kirby, K. S. (1967). *Methods Cancer Res.* **3,** 1.

Klimek, M. (1965). *Neoplasma* **12,** 559.

Klimek, M. (1966). *Photochem. Photobiol.* **5,** 603.

Knowlton, N. P., Jr., and Hempelmann, L. H. (1949). *J. Cell. Comp. Physiol.* **33,** 73.

Koch, J., and Stokstrad, E. L. R. (1967). *Eur. J. Biochem.* **3,** 1.

Kouri, R. E., and Coggin, J. H. (1968). *Proc. Soc. Exp. Biol. Med.* **129,** 609.

Krueger, R. G., and McCarthy, B. J. (1970). *Biochem. Biophys. Res. Comm.* **41,** 944.

Lajtha, L. G. (1968). *Radiat. Res.* **33,** 659.

Lajtha, L. G., Oliver, R., and Ellis, F. (1954). *Brit. J. Cancer* **8,** 367.

Lajtha, L. G., Oliver, R. and Kumatori, T., and Ellis, F. (1958a). *Radiat. Res.* **8,** 1.

Lajtha, L. G., Oliver, R., Berry, R., and Noyes, W. D. (1958b). *Nature* (*London*) **182,** 1788.

Landy, M., and Braun, W., eds. (1969). "Immunological Tolerance." Academic Press, New York.

Langham, W., Woodward, K. T., Rothermel, S. M., Harris, P. S., Lushbaugh, C. C., and Storer, J. B. (1956). *Radiat. Res.* **5,** 404.

Lea, D. E. (1955). "Actions of Radiations on Living Cells." Cambridge Univ. Press, London and New York.

Lehnert, S., and Okada, S. (1963). *Nature* (*London*) **199,** 1108.

Lehnert, S., and Okada, S. (1966). *Int. J. Radiat. Biol.* **10,** 601.

Lesher, S., and Vogel, H. H. (1958). *Radiat. Res.* **9,** 560.

Little, J. R., and Donahue, H. A. (1970). *Proc. Nat. Acad. Sci. U. S.* **67,** 1299.

Littlefield, J. W. (1966). *Biochim. Biophys. Acta* **114,** 398.

McCarthy, B. J., and Hoyer, B. H. (1964). *Proc. Nat. Acad. Sci. U. S.* **52,** 915.

McCulloch, E. A., and Till, J. E. (1960). *Radiat. Res.* **13,** 115.

McGrath, R. A. (1960). *Int. J. Radiat. Biol.* **2**, 177.
McGrath, R. A., and Congdon, C. C. (1959). *Int. J. Radiat. Biol.* **1**, 80.
McKusick, V. A. (1966). "Mendelian Inheritance in Man." Johns Hopkins Press, Baltimore, Maryland.
Mak, S., and Till, J. E. (1963). *Radiat. Res.* **20**, 600.
Makinodan, T. (1966). *In* "English Encyclopedia of Medical Radiology. Part II. Radiation Biology" (A. Zuppinger, ed.), p. 303. Springer-Verlag, Berlin and New York.
Makinodan, T., and Price, G. B. (1971). *In* "Transplantation" (J. S. Najarian, and R. L. Simmons, eds.), Chapter 5, Lea & Febiger, Phila. Penna.
Manoukhine, I. I. (1913). *C. R. Soc. Biol.* **74**, 1221.
Marin, G., and Littlefield, J. W. (1968). *J. Virol.* **2**, 69.
Marin, G., and MacPherson, I. (1969). *J. Virol.* **3**, 146.
Martin, G. S. (1970). *Nature (London)* **227**, 1021.
Mayfield, J. E., Holmes, D. S., and Bonner, J. (1971). *Biophys. Soc. Abstr.* **15**, 151a.
Miller, J. F. A. P. (1964). *Science* **144**, 1544.
Millstein, C., and Monro, A. J. (1970). *Annu. Rev. Microbiol.* **24**, 335.
Mitchell, G. F., and Miller, J. F. A. P. (1968). *J. Exp. Med.* **128**, 821.
Mitchison, J. M. (1969). *In* "The Cell Cycle" (G. M. Padilla, G. L. Whitson, and I. L. Cameron, eds.), p. 361. Academic Press, New York.
Mitchison, N. A. (1969). *In* "Immunological Tolerance" (M. Landy, and W. Braun, eds.), p. 113. Academic Press, New York.
Montagna, W., and Wilson, J. W. (1955). *J. Nat. Cancer Inst.* **15**, 1703.
Mosier, D. E., and Coppleson, L. (1968). *Proc. Nat. Acad. Sci. U. S.* **61**, 542.
Mosier, D. E., Fitch, F. W., Rowley, D. A., and Davies, A. J. S. (1970). *Nature (London)* **225**, 276.
Muller, H. J. (1927). *Science* **66**, 84.
Nowell, P. C. (1965). *Blood* **26**, 798.
Nowell, P. C. (1967). *In* "Human Radiation Cytogenetics" (H. J. Evans, W. M. Court Brown, and A. S. McLean, eds.), p. 99. North-Holland Publ., Amsterdam.
Nowell, P. C., and Cole, L. J. (1963). *Science* **141**, 524.
Okada, S. (1970). "Radiation Biochemistry" (K. I. Altman, G. B. Gerber, and S. Okada, eds.), Vol. 1, p. 189. Academic Press, New York.
Okada, S., and Hempelmann, L. H. (1959). *Int. J. Radiat. Biol.* **1**, 305.
Ord, M. G., and Stocken, L. A. (1958). *Nature (London)* **182**, 1787.
Painter, R. B., and Hughes, W. L. (1961). *Ann. N. Y. Acad. Sci.* **95**, 960.
Painter, R. B., and Robertson, J. S. (1959). *Radiat. Res.* **11**, 206.
Pearsall, N. N., and Weiser, R. S. (1970). "The Macrophage." Lea & Febiger, Philadelphia, Pennsylvania.
Perkins, E. H., Nettesheim, P., and Makinodan, T. (1966). *J. Reticuloendothel. Soc.* **3**, 71.
Prescott, D. M., and Bender, M. A. (1962). *Exp. Cell Res.* **26**, 260.
Pribnow, J. F., and Silverman, M. S. (1967). *J. Immunol.* **98**, 225.
Puck, T. T., and Steffen, J. (1963). *Biophys. J.* **3**, 379.
Puck, T. T., and Yamada, M. (1962). *Radiat. Res.* **16**, 589.
Puck, T. T., Sanders, P., and Petersen, D. (1964). *Biophys. J.* **4**, 441.
Quastler, H. (1956). *Radiat. Res.* **4**, 303.
Quastler, H., and Hampton, J. C. (1962). *Radiat. Res.* **17**, 914.
Rasmussen, R. E., and Painter, R. B. (1966). *J. Cell Biol.* **29**, 11.
Regan, J. D., Trosko, J. E., and Carrier, W. L. (1968). *Biophys. J.* **8**, 319.

Reif, A. E., and Allen, J. M. (1966). *Nature (London)* **209,** 521.

Riggsby, W. S., Jones, N. D., and Godden, W. R. (1966). USAF Tech. Rep. AFWL-TR-65-112.

Robbins, E., and Borum, T. (1967). *Proc. Nat. Acad. Sci. U. S.* **57,** 409.

Robbins, E., and Gonatas, N. K. (1964). *J. Cell Biol.* **21,** 429.

Robbins, E., Jentzoch, G., and Micali, A. (1968). *J. Cell Biol.* **36,** 229.

Roche, J. G., Rosenau, W., and Goldberg, M. L. (1969). *Proc. Soc. Exp. Biol. Med.* **131,** 465.

Rook, A. J., Wilkinson, D. S., and Ebling, F. J. G. (1968). "Textbook of Dermatology," Vol. 1, p. 62. Davis, Philadelphia, Pennsylvania.

Roseman, J. (1969). Science **165,** 1125.

Roth, J. S. (1967). *Methods Cancer Res.* **3,** 154.

Rowe, D. S. (1970). *Nature (London)* **228,** 509.

Rowley, D. A., Fitch, F. W., Mosier, D. E., Solliday, S., Coppleson, L. W., and Brown, B. W. (1968). *J. Exp. Med.* **127,** 983.

Rustad, R. C. (1960). *Exp. Cell Res.* **21,** 596.

Rytömaa, T., and Kiviniemi, K. (1968a). *Cell Tissue Kinet.* **1,** 329.

Rytömaa, T., and Kiviniemi, K. (1968b). *Cell Tissue Kinet.* **1,** 341.

Rytömaa, T., and Kiviniemi, K. (1968c). *Eur. J. Cancer* **4,** 595.

Sado, T. (1969). *Int. J. Radiat. Biol.* **15,** 1.

Santos, G. W., and Owens, A. H. (1966). *Bull. Johns Hopkins Hosp.* **118,** 109.

Sarkar, N., Devi, A., and Hempelmann, L. H. (1961). *Nature (London)* **192,** 179.

Scharff, M. D., and Robbins, E. (1965). *Nature (London)* **208,** 464.

Scharff, M. D., and Robbins, E. (1966). *Science* **151,** 992.

Scherrer, D., Spohr, G., Nicole, G., Granboulan, N., Morel, C., Grosclaude, J., and Chezzi, C. (1970). *Cold Spring Harbor Symp. Quant. Biol.* **35,** 539.

Scolnick, E. M., Aaronson, S. A., Todaro, G. J., and Parks, W. P. (1971). *Nature (London)* **229,** 318.

Seguira, J. H., Ingram, J. T., and Brain, R. T. (1911). "Diseases of the Skin," Macmillan, New York.

Sercarz, E., and Coons, A. H. (1962). *In* "Mechanisms of Immunological Tolerance" (M. Hašek, A. Lengerová, and M. Vojtíšková, eds.), p. 73. Publ. House Czech. Acad. Sci., Prague.

Setlow, R. B., Regan, J. D., German, J., and Carrier, W. L. (1969). *Proc. Nat. Acad. Sci. U. S.* **64,** 1035.

Shearer, R., and McCarthy, B. J. (1967). *Biochemistry* **6,** 283.

Shepard, D. C. (1965). *Exp. Cell Res.* **38,** 570.

Sherman, F. G., and Quastler, H. (1960). *Exp. Cell Res.* **19,** 343.

Silverman, M. S., Greenman, V., Chin, P., and Bond, V. P. (1958). *Radiat. Res.* **8,** 128.

Simić, M. M., and Petrović, M. Ž. (1971). *Nature (London)* **229,** 263.

Simić, M. M., Šlijivć, V. S., Petrović, M. Ž., and Ćirković, D. M. (1965). *Bull. Boris Kidric Inst. Nucl. Sci.* **16,** Suppl. 1, 1.

Sinclair, W. K. (1968). *Radiat. Res.* **33,** 620.

Smets, L. A., and Dewaide, H. (1966). *Naturwissenschaften* **53,** 382.

Smith, L. H., and Congdon, C. C. (1960). *In* "Radiation Protection and Recovery" (A. Hollaender, ed.), p. 242. Pergamon, Oxford.

Smith, R. T., Miescher, P. A., and Good, R. A., eds. (1966). "Phylogeny of Immunity." Univ. of Florida Press, Gainesville.

Soeiro, R., and Darnell, J. E. (1969). *J. Mol. Biol.* **44,** 551.
Stanners, C. P., and Till, J. E. (1960). *Biochim. Biophys. Acta* **37,** 406.
Šterzl, J. (1967). *Cold Spring Harbor Symp. Quant. Biol.* **32,** 493.
Steward, D. L., Schaeffer, J. R., and Humphrey, R. M. (1968). *Science* **161,** 791.
Stoker, M. (1963). *Nature (London)* **200,** 756.
Stonehill, G. H., and Bendich, A. (1970). *Nature (London)* **228,** 370.
Stoner, R. D., and Hale, W. M. (1962). *In* "Effects of Ionizing Radiation on Immune Processes" (C. A. Leone, ed.), p. 183. Gordon and Breach, New York.
St. Pierre, R. L., and Ackerman, G. A. (1965). *Science* **147,** 1307.
Stubblefield, E., and Murphree, S. (1968). *Exp. Cell Res.* **48,** 652.
Sullivan, M. F., Marks, S., Hackett, P. L., and Thompson, R. C. (1959). *Radiat. Res.* **11,** 653.
Szakal, A. K., and Hanna, M. G. (1968). *Exp. Mol. Pathol.* **8,** 75.
Taliaferro, W. H., Taliaferro, L. G., and Jaroslow, B. N. (1964). "Radiation and Immune Mechanisms" Academic Press, New York.
Talmage, D. W., Radovich, J., and Hemmingson, H. (1970). *Advan. Immunol.* **12,** 271.
Taylor, E. W. (1965). *Exp. Cell Res.* **40,** 316.
Taylor, R. B. (1969). *Transplant. Rev.* **1,** 114.
Terasima, T., and Tolmach, L. J. (1963a). *Exp. Cell Res.* **30,** 334.
Terasima, T., and Tolmach, L. J. (1963b). *Biophys. J.* **3,** 11.
Terasima, T., and Yasukawa, M. (1966). *Exp. Cell Res.* **44,** 669.
Thomas, C. A., Jr., Hamkalo, B. A., Misra, D. N., and Lee, C. S. (1970). *J. Mol. Biol.* **51,** 621.
Till, J. E., and McCulloch, E. A. (1961). *Radiat. Res.* **14,** 213.
Till, J. E., and McCulloch, E. A. (1964). *Radiat. Res.* **22,** 383.
Tobey, R. A., Petersen, D. F., Anderson, E. C., and Puck, T. T. (1966). *Biophys. J.* **6,** 567.
Tori, G., and Gasbarrini, G. (1963). *Radiol. Clin.* **32,** 47.
Trosko, J. E., and Kasschau, M. R. (1967). *Photochem. Photobiol.* **6,** 215.
Trosko, J. E., Chu, E. H. Y., and Carrier, W. L. (1965). *Radiat. Res.* **24,** 667.
Uhr, J. W. (1964). *Science* **145,** 457.
Upton, A. C. (1968). *Methods Cancer Res.* **4,** 54.
Valencia, J. I., and DeLozzio, C. B. (1962). *Radiat. Res.* **16,** 18.
Van Lancker, J. L. (1960). *Biochim. Biophys. Acta* **45,** 57.
Walters, R. A., and Petersen, D. F. (1968a). *Biophys. J.* **8,** 1475.
Walters, R. A., and Petersen, D. F. (1968b). *Biophys. J.* **8,** 1487.
Watanabe, I., and Okada, S. (1966). *Radiat. Res.* **27,** 290.
Watanabe, I., and Okada, S. (1967). *J. Cell Biol.* **32,** 309
Watanabe, I., and Okada, S. (1968). *Radiat. Res.* **35,** 202.
Watson, J. D. (1965). "The Molecular Biology of the Gene." Benjamin, New York.
Welling, W., and Cohen, J. A. (1960). *Biochim. Biophys. Acta* **42,** 181.
Whitehouse, H. L. K. (1967). *J. Cell Sci.* **2,** 9.
Whitmore, G. F., Stanners, C. P., Till, J. E., and Gulyas, S. (1961). *Biochim. Biophys. Acta* **47,** 66.
Williams, R. B., Jr., Toal, J. N., White, J., and Carpenter, H. M. (1958). *J. Nat. Cancer Inst.* **21,** 17.
Wolff, S. (1960). *Amer. Natur.* **94,** 85.
Wolff, S., and Luippold, H. E. (1955). *In* "Progress in Radiobiology" (J. S. Mitchell, B. E. Holmes, and C. L. Smith, eds.), p. 217. Oliver & Boyd, Edinburgh.

Wu, T. E., and Kabat, E. A. (1970). *J. Exp Med.* **132,** 211.
Xeros, N. (1962). *Nature (London)* **194,** 682.
Yamada, M., and Puck, T. T. (1961). *Proc. Nat. Acad. Sci. U. S.* **47,** 1181.
Yu, C. K., and Sinclair, W. K. (1967). *J. Nat. Cancer Inst.* **39,** 619.

Chapter 7

Ionizing Radiation Effects on Higher Plants

ALAN H. HABER

I. Introduction

Higher plants provide exceptionally favorable material for many basic studies in radiobiology. Actions of visible light—photosynthesis, photomorphogenesis, and phototropism (Seliger and McElroy, 1965)—are studied intensively and, to a lesser extent, the effects of ultraviolet radiation (Lockhart and Franzgrote, 1961). One aspect of the usefulness of plant systems involves the feasibility of experimentally separating growth, cell division, and senescence from one another. This chapter will stress such favorable properties of plants for research in ionizing radiation biology. A review of the complementary point of view—that ionizing radiobiology is useful for research in plant physiology—is published elsewhere (Haber, 1968).

II. Actions of Radiation on Chromosomes and Mitosis

At the cytological level ionizing radiation produces two readily discernible effects: chromosome aberrations and mitotic inhibition.

A. Radiation-Induced Chromosomal Aberrations

During the normal mitotic cycle in unirradiated cells, individual chromosomes seem to maintain their integrity without breakage or rearrangements. After irradiation, however, chromosomes or strands of chromosomes can break at various points along their lengths. At the subsequent telophase an acentric fragment (i.e., piece of a chromosome lacking a centromere) is unable to move to either of the newly constituted daughter nuclei. Such chromosome fragments remain outside the nuclei and their genes are lost to further progeny of the original, irradiated mother cell. These fragments can be detected as Feulgen-staining "micronuclei" in the cytoplasm of daughter cells formed by division of the irradiated cell. This type of deletion is the simplest type of chromosomal aberration that leads to gene deficiencies in daughter nuclei produced by mitosis after irradiation. Another type of chromosomal abnormality results from rejoining of two different broken chromosomes in such a manner that the newly formed structure has two centromeres. If the two centromeres happen to move to opposite poles at anaphase, then the chromosomal

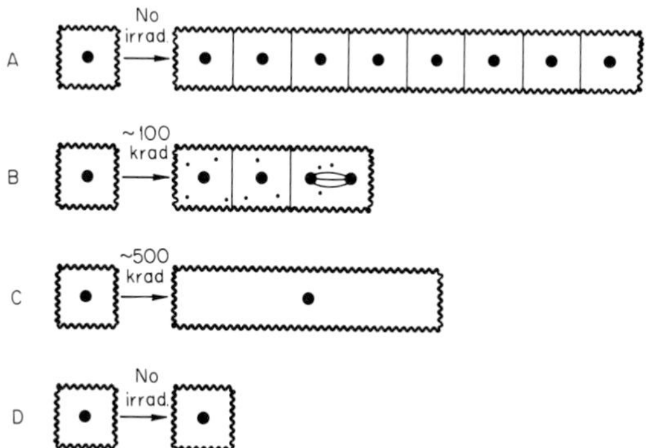

Fig. 1. Diagrammatic representation of some of the biological systems discussed. A. Normal growth with cell divisions in unirradiated plants. B. Growth of seedling with nuclear imbalance resulting from division of cells with chromosomal aberrations. C. Growth of seedling (the "γ-plantlet") without cell division after massive γ-irradiation of the grain before sowing. D. Apical wheat leaf sections which undergo normal physiological senescence in the absence of both cell division and growth.

material becomes stretched between them to produce a "chromosome bridge" across the cytoplasm. The genes on this extranuclear chromosome bridge, like those on the micronuclei previously discussed, are lost to the progeny of the irradiated cell. Micronuclei and bridges are diagrammatically represented in Fig. 1B. There are many more types of chromosomal rearrangements that occur after irradiation; for a more complete, elementary discussion the reader is referred to Swanson (1957). For radiation-induced cell lethality, the importance of the chromosomal abnormalities lies in the production of nuclear imbalance, usually occurring as deficiencies in large numbers of genes. Such nuclear imbalance cannot occur, however, unless the nucleus divides after irradiation.

B. Mitotic Inhibition

In general, cell division seems to be more readily inhibited by irradiation than is cell growth. In some cases, complete prevention of mitosis by massive seed irradiation does not prevent a growth that is remarkably normal in many respects. Seedlings growing without cell division after massive γ-irradiation of dry seeds are called "γ-plantlets" (Haber, 1968). The mechanism for this radiation-induced mitotic inhibition is unknown. The mitotic inhibition may or may not be accompanied by complete inhibition of DNA synthesis (Haber and Foard, 1964a; Haber *et al.*,

1969a). As will be discussed in Section II,C and in Section VI, radiation-induced mitotic inhibition in itself does not necessarily imply lethality.

C. Interaction between Chromosomal Breakage and Mitosis in Producing Nuclear Imbalance

As discussed in Section II,A, chromosome breakage leads to nuclear imbalance when it is followed by mitosis (cf. Figs. 1A and 1B). With an increasing radiation dose, the extent of chromosome breakage and rearrangements increases, but the mitotic frequency decreases (cf. Figs. 1A, 1B, and 1C). After receiving very high doses, certain seeds germinate with little or no mitosis (Fig. 1C); these are the γ-plantlets discussed in Section II,B. Consequently, there is little or no nuclear imbalance (cf. Figs. 1C and 1B). This interaction of chromosomal breakage and mitosis in producing nuclear imbalance was first reported by Schwartz and Bay (1956) for corn seedlings from irradiated grains and has since been extended to other cereals (Haber, 1968). In summary, chromosome breakage and rearrangements lead to cell lethality only when they are followed by mitosis.

III. Actions of Radiation on Cell Growth

A. Effects Mediated by Mitotic Inhibition

Conceptually and geometrically, cell growth and cell division are different processes. When such occurrences as changes in volume of intercellular spaces, sloughing-off of cells, and tissue shrinkage, are neglected, the net expansion of an organ or organism is identical to the total amount of cell expansion (Haber and Foard, 1964b). Biologically, however, cell expansion and cell division are causally interrelated. A variety of experiments and observations exposed a fallacy in the often accepted manner of interpreting changes in organ size as being due to changes in cell size and changes in cell number. The presence or absence of cell divisions in growing seedlings does not causally affect the rate of concurrent growth; the absence of cell divisions, however, will subsequently limit the ultimate extent of future growth (Haber, 1962, 1963; Haber and Foard, 1963; Foard *et al.*, 1965). Accordingly, radiation-induced mitotic inhibition in itself (i.e., apart from any other action of radiation) can limit the ultimate extent of growth. Even complete mitotic inhibition, however, need not in itself produce a decrease in concurrent growth rate. This idea will be further developed in Section III,C.

B. Effects Mediated by Nuclear Imbalance

That nuclear imbalance resulting from division of cells with chromosomal aberrations can be inhibitory to growth is illustrated in Fig. 1. After mitosis in cereal seeds has been blocked by high doses of X- or γ-radiation (Fig. 1C), it is often noted that they actually grow better than seedlings from seeds given lower doses, after which low dose, incomplete mitotic inhibition results in nuclear imbalance (Fig. 1B; Haber, 1968). Since the radiation dose is less as shown in Fig. 1B than in Fig. 1C, the decreased growth in the former must be due to the nuclear imbalance that results from division of cells that have chromosomal aberrations, diagrammatically represented as micronuclei and chromosome bridges (see Section II,A).

C. Effects on Cell Growth per Se

Systems are available for studying radiation actions on cell growth uncomplicated by any actions of radiation either in producing mitotic inhibition (see Section III,A) or in producing nuclear imbalance that results from the division of cells with chromosomal aberrations (see Section III,B). The compound 5-fluorodeoxyuridine (FUdR) inhibits DNA synthesis but has little or no inhibitory action on growth apart from its inhibition of DNA synthesis and mitosis (Haber and Triplett, 1970). The growth of wheat seedlings growing with DNA synthesis and mitoses can be greatly inhibited by FUdR concentrations as low as 0.0001 M. The growth of γ-plantlets, the seedlings growing with inhibited DNA synthesis and without cell division after 500 to 800 krads of γ-irradiation of the dry seeds, is unaffected by FUdR concentrations as high as 0.1 M. Comparison of the growth curves of both γ-plantlet and unirradiated (control) seedlings treated with various concentrations of FUdR leads to the conclusion that the γ-radiation used to produce γ-plantlets has little or no inhibitory effect on growth by mechanisms other than the inhibition of DNA synthesis and mitosis (Haber and Schwarz, 1971). That this γ-radiation, which produces γ-plantlets, does not inhibit growth by mechanisms other than mitotic inhibition is also indicated by similar experiments with the mitotic inhibitor maleic hydrazide. When an unirradiated seed germinates in maleic hydrazide at concentrations that prevent cell division, the seedling grows to the same size as a γ-plantlet, which is unaffected by the maleic hydrazide treatment (Haber and White, 1960). Two additional indications that cell growth in γ-plantlets is inhibited entirely or almost entirely by the mitotic inhibition are (a) in embryo culture the initial growth rates of γ-plantlet leaves growing without cell division are the same as unirradiated leaves growing with many cell divisions (see Fig. 2 of Long and Haber, 1965), and (b) among those structures (e.g., coleorhiza, epiblast, root hairs), which in unirradiated controls have no cell division (and thus no cell division for the irradiation to inhibit), the final size attained is the same in γ-plantlets and unirradiated controls (Haber and Foard, 1964a). Consequently, we conclude that, after the massive seed irradiation (0.5 to 0.8 Mrad of γ-rays) that prevents cell division during germination of wheat, the decreased growth observed relative to unirradiated seedlings can be attributed largely or entirely to consequences of the mitotic inhibition (as discussed in Section III,A), and not to other effects of the γ-radiation.

Irradiation of growing γ-plantlets as seedlings provides a system for studying effects of seedling irradiation on cell growth per se, since the earlier seed irradiation has already prevented mitosis. The seedling irradiation cannot inhibit growth by inducing any mitotic inhibition (as described in Section III,A), because there is no mitosis to inhibit; neither can the seedling irradiation inhibit growth by producing nuclear imbalance resulting from division of cells with chromosomal aberrations (as described in Section III,B), because there is no mitosis. Such studies indicate that seedling irradiation is more effective than dry-seed irradiation for inhibiting growth. It is not possible to describe numerically the increased sensitivity, because the radiation is given at different times during development. Within the range of dry-seed radiation doses sufficient to produce γ-plantlets, 300 krads of additional seed irradiation is necessary to give an inhibition of cell growth per se comparable to the inhibition produced by only about 50 krads given to the growing γ-plantlet seedlings (Haber *et al.*, 1969c).

IV. Actions of Radiation on Senescence and Death

Senescence is much more difficult to define in plants than in animals because its end point, death, cannot be defined as precisely in plants. Nevertheless, much is known of plant senescence, which in some ways is more easily studied than animal senescence because of the ease of environmental control (Woolhouse, 1967).

A. Effects Mediated by Nuclear Imbalance

Nuclear imbalance, as first described in Section II,C, and as discussed in relation to growth inhibition in Section III,B, can also promote senescence. In seedlings growing from irradiated cereal grains, necrosis is observed after doses that produce nuclear imbalance resulting from division of cells with chromosomal aberrations (Fig. 1B), but not after higher doses that do not result in such nuclear imbalance (Fig. 1A; Schwartz and Bay, 1956).

In animals, low doses of radiation appear to accelerate senescence; much if not all of this accelerated senescence may be attributed to radiation-induced somatic mutations (Curtis, 1967). Compared to animal senescence, plant senescence seems to be more strictly controlled by genetic programming and by environmental conditions; perhaps this is why low doses of radiation have not been implicated in overall shortening of the life-span of plants.

B. Effects Mediated by Growth Inhibition

Apart from any possible radiation-induced acceleration of senescence caused by nuclear imbalance (see Section IV,A) or caused by direct actions on senescence per se (see Section IV,C), radiation-induced limitation of growth in itself seems capable of causing an earlier senescence. This concept of a premature senescence resulting solely from an earlier cessation of growth, irrespective of the cause of this earlier cessation, has been reviewed by Kohn (1965). It can perhaps be illustrated also by results of comparative studies of senescence in γ-plantlets and unirradiated plants. In the leaf, senescence can be followed as the slow loss of chlorophyll, a loss that also parallels the loss of leaf protein and nucleic acids as well as the disintegration of chloroplasts and other organelles (Shaw *et al.*, 1965; Shaw and Manocha, 1965; Walne and Haber, 1968). Due primarily to the mitotic inhibition (see Sections III,A and III,C), the first foliage leaf stops growing in the γ-plantlet before it stops in the unirradiated seedling. The time course of senescence after cessation of its growth, however, is very similar in γ-plantlets and unirradiated seedlings (Foard and Haber, 1970). In cortical parenchyma of roots, senescence can be followed as the loss of capacity to incorporate radioactive uridine into insoluble RNA. In the epidermis of roots, senescence can be followed as cessation of protoplasmic streaming in root hairs. By both these criteria in roots, senescence occurs within a day after cessation of growth in any given region of the root (Foard and Haber, 1970). We conclude that one component of the acceleration of senescence by irradiation is the earlier cessation of growth.

C. Effects on Senescence per Se

The apical portion of the normal, typical cereal leaf provides a system for studying senescence per se, i.e., senescence in the absence of both cell division and growth. This portion of the cereal leaf, which performs its normal functions in photosynthetic metabolism in the absence of both cell division and growth, slowly senesces in a manner diagrammatically represented in Fig. 1D. Any effects of ionizing radiation on its senescence cannot be attributed to nuclear imbalance resulting from division of cells with chromosomal aberrations, because there is no cell division even in unirradiated controls. Nor can any effects of ionizing radiation on such senescence be attributed to growth inhibition or mitotic inhibition, since there is no cell growth and no cell division to inhibit, even in unirradiated controls. In studies of three varieties of wheat, γ-radiation doses as high as 100 to 400 krads were found to be completely without detectable effect on senescence as measured by rate of chlorophyll loss (Haber and Walne, 1968). In contrast, senescence in these systems is very strikingly regulated, both positively and negatively, by treatment with low concentrations of various chemicals, by temperature, or by illumination with low intensities of light (Foard and Haber, 1970; Haber and Walne, 1968; Haber *et al.*, 1969b). Thus a senescence that is easily regulated by many physical and chemical treatments is unaffected by remarkably high doses of ionizing radiations, provided there is an absence of the typical radiation effects of (a) nuclear imbalance resulting from division of cells with chromosomal aberrations, (b) radiation-induced mitotic inhibition, (c) radiation-induced growth inhibition, and (d) amplification of somatic mutations by cell division.

D. Separation of Death from Physiological Senescence

In the experiments described in Section IV,C, we could find no γ-radiation doses that accelerated senescence without direct killing of leaf tissue (Haber and Walne, 1968). These lethal doses ranged from 250 to 800 krads. Acceleration of senescence in green tissue is indicated by an acceleration of the slow loss of chlorophyll, whether in light or in darkness, relative to controls that received no ionizing irradiation. In contrast, death is indicated by rapid chlorophyll loss in the light but inhibition of chlorophyll loss in the dark (Haber and Walne, 1968). Therefore, in the apical cereal-leaf system, in which there is neither cell division nor growth, direct cell killing can be experimentally separated from acceleration of senescence.

V. Law of Bergonié and Tribondeau

Up to this point we have dealt with several aspects of the relations between cell division and radiation injury. The "Law of Bergonié and Tribondeau" has been accepted as a generalization concerning the greater effectiveness of ionizing radiation on dividing over nondividing cells. This law, however, is less a precisely stated scientific law subject to experimental test than it is a general point of view reflecting several aspects:

1. Radiation effects are more easily seen in dividing than in nondividing cells.
2. Dividing cells have greater intrinsic radiosensitivity.
3. After irradiation, cell divisions increase damage.

The first aspect is generally true but essentially trivial, because some of the most characteristic cytological radiation effects—mitotic inhibition, gross genic imbalance, and multipolar spindles—could not possibly occur in cells not undergoing mitosis. Moreover, other effects, such as chromosome breakage or somatic mutation, can occur in both dividing and nondividing cells but can be detected only in dividing cells.

The second aspect is a statement contradicted by experiments with another favorable plant system: homoblastic leaf development in tobacco. The rate of cell division decreases as tobacco leaves grow, and older leaves grow without any cell divisions. Discs were cut from leaves having differing degrees of cell division at various developmental stages. In the chlorenchyma, the induced susceptibility to photodestruction of chloroplasts was used to measure sensitivity to γ-radiation as a function of the rate of cell division. The same biological effect can thereby be studied both in dividing and in nondividing tissues of the same morphological and physiological cell types. Since the criterion used involved death very soon after high doses of irradiation, postirradiation cell divisions had little or no effect on the radiosensitivity that was studied. In this system, radiosensitivities were approximately the same, irrespective of the extent of cell division occurring at the time of irradiation (Haber and Rothstein, 1969).

The third aspect is neither trivial nor false. It is a valid generalization we have already dealt with in connection with the role played by postirradiation cell divisions in producing nuclear imbalance (see Section II,C). Postirradiation cell divisions also amplify somatic mutations by perpetuating cell lineages from individual, mutated cells. Thus, extensive cell division after irradiation will result in further damage.

These considerations do not contradict any of the observations pertinent to the Law of Bergonié and Tribondeau, because other correlations of apparent radiation sensitivity with mitotic activity have entailed observations of different morphological cell types involving different effects or have used criteria of radiation effects inapplicable to nondividing cells. Within the same type of tissue, there undoubtedly are physiological differences associated with the presence or absence of cell division. The results do suggest, however, that such physiological differences do not significantly alter the intrinsic radiosensitivity of the whole cell (Haber and Rothstein, 1969).

VI. Separation of "Genetic" and "Physiological" Lethality

In Section III,C there was a discussion of the normalcy of cell growth in γ-plantlets. These seedlings must certainly be considered dead in the genetic sense, however, since there can be neither reproduction of the plant, nor even reproduction of its cells. Nevertheless, the γ-plantlet is surprisingly normal in many other respects in addition to growth. Among these additional similarities to normal seedlings are polarization of growth, net protein synthesis as well as synthesis of a complex of specific enzymes, development of the photosynthetic apparatus, organelle development, nutritional requirements, cell and tissue maturation, correlative organ growth, chemical regulation of growth, and certain aspects of nuclear behavior and function despite the absence of DNA synthesis and mitosis (for review, see Haber, 1968). The γ-plantlet even resembles normal seedlings in its response to seedling irradiation (Haber *et al.*, 1969c).

Because of the unquestioned genetic lethality in γ-plantlets and because of the very high doses of radiation that are used, the question inevitably arises: How long do γ-plantlets live? This question was put in a form amenable to experiments comparing senescence of γ-plantlets with unirradiated controls. For this comparison we studied leaf senescence by examining the time course and chemical regulation of the chlorophyll loss that reflects overall senescence of wheat leaves (Shaw *et al.*, 1965; Shaw and Manocha, 1965) as well as loss of chloroplast ultrastructure (Haber and Walne, 1968; Walne and Haber, 1968). We explored root senescence by studying the capacity of roots of different ages to incorporate ^{3}H-uridine into insoluble RNA in different morphological regions. We found that the slow chlorophyll loss in the γ-plantlet leaf resembles that in the unirradiated control with respect to time course and chemical regulation. In both γ-plantlet and normal roots, protoplasmic streaming occurs in all growing root hairs, and synthesis of insoluble RNA is detected in all cells before they stop growing. In root hairs of both γ-plantlets and unirradiated controls, protoplasmic streaming stops within a day after cessation of their growth. In a given region of roots, detectable synthesis of insoluble RNA stops within a day of cessation of growth in that region. An exception is the γ-plantlet root apex, together with some pericycle tissue extending somewhat more basally, which may retain the capacity for detectable synthesis of insoluble RNA for up to 2.5 weeks after cessation of its growth. This contrasts sharply with other regions of the γ-plantlet root and all regions of unirradiated roots, in which detectable synthesis of insoluble RNA stops within a day of cessation of their growth. In γ-plantlet roots the regions in which synthesis of insoluble RNA persists, uncoupled from growth, might be considered to correspond histologically to regions in which continued development of the main root and the initiation of lateral roots is normally perpetuated (Foard and Haber, 1970). These normal aspects of senescence in γ-plantlets indicate that doses of 0.5 Mrads or more given to dry wheat grains are not lethal in the physiological sense despite their lethality in the genetic and proliferative senses. Thus the γ-plantlet provides a rather clear experimental separation of genetic from physiological aspects of radiation-induced lethality. The absence of nuclear imbalance, in addition to other characteristics of the dry-seed irradiation, may also underlie the uniqueness of the γ-plantlet system (Foard and Haber, 1970).

VII. Summary

The causal interrelations among the various aspects of radiation injury in reducing growth and promoting senescence and death are diagrammatically summarized in Fig. 2. Much of the growth inhibition is a consequence of mitotic inhibition, although systems are available for studying inhibition of cell growth per se in nondividing systems. Radiation seems to promote senescence by means of nuclear imbalance and growth inhibition, but not significantly by accelerating senescence in absence of growth, cell division, and nuclear imbalance. The role of cell division

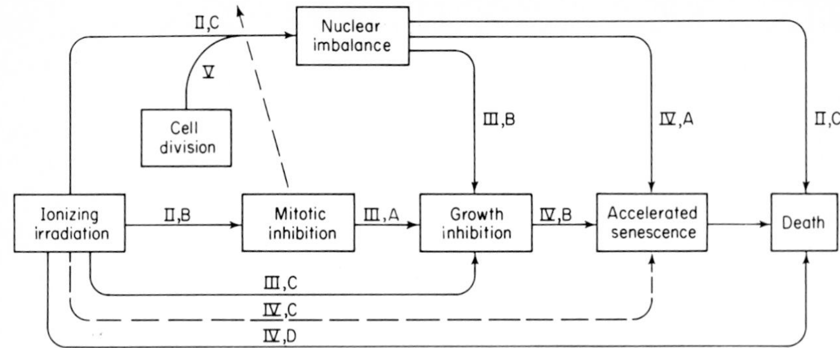

Fig. 2. Diagrammatic summary of causal interrelations of various processes involved in growth inhibition and accelerated senescence. Roman numerals and letters refer to sections of the text in which the causal relation is discussed.

in radiation injury is not to increase intrinsic radiosensitivity, but to amplify the damage after irradiation by producing a nuclear imbalance that results from chromosome breakage and perhaps also by perpetuating somatic mutations in the progeny of mutated cells. In the absence of the many deleterious consequences of postirradiation cell divisions, seedlings showing complete lethality in the genetic sense can be surprisingly normal in the physiological sense.

Acknowledgments

This research was sponsored by the United States Atomic Energy Commission under contract with the Union Carbide Corporation.

References

Curtis, H. J. (1967). *Symp. Soc. Exp. Biol.* **21,** 51–63.

Foard, D. E., and Haber, A. H. (1970). *Radiat. Res.* **42,** 372–380.

Foard, D. E., Haber, A. H., and Fishman, T. N. (1965). *Amer. J. Bot.* **52,** 580–590.

Haber, A. H. (1962). *Amer. J. Bot.* **49,** 583–589.

Haber, A. H. (1963). *In* "Plant Tissue Culture and Morphogenesis" (J. C. O'Kelley, ed.), pp. 79–88. Scholar's Library, New York.

Haber, A. H. (1968). *Annu. Rev. Plant Physiol.* **19,** 463–489.

Haber, A. H., and Foard, D. E. (1963). *Amer. J. Bot.* **50,** 937–944.

Haber, A. H., and Foard, D. E. (1964a). *Amer. J. Bot.* **51,** 151–159.

Haber, A. H., and Foard, D. E. (1964b). *Colloq. Int. Cent. Nat. Rech. Sci.* **123,** 491–503.

Haber, A. H., and Rothstein, B. E. (1969). *Science* **163,** 1338–1339.

Haber, A. H., and Schwarz, O. J. (1970). *Plant Physiol.* **45,** Suppl., 31.

Haber, A. H., and Triplett, L. L. (1970). *ASB Bull.* **17,** 45. (The official quarterly publication of the Association of Southeastern Biologists.)
Haber, A. H., and Walne, P. L. (1968). *Radiat. Bot.* **8,** 389–397.
Haber, A. H., and White, J. D. (1960). *Plant Physiol.* **35,** 495–499.
Haber, A. H., Foard, D. E., and Perdue, S. W. (1969a). *Plant Physiol.* **44,** 463–467.
Haber, A. H., Thompson, P. J., Walne, P. L., and Triplett, L. L. (1969b). *Plant Physiol.* **44,** 1619–1628.
Haber, A. H., Foard, D. E., and Triplett, L. L. (1969c). *Radiat. Bot.* **9,** 473–479.
Kohn, R. R. (1965). *In* "Reproduction: Molecular, Subcellular and Cellular" (M. Locke, ed.), pp. 291–324. Academic Press, New York.
Lockhart, J., and Franzgrote, U. B. (1961). *In* "Handbuch der Pflanzenphysiologie" (W. Ruhland, ed.), Vol. **16,** pp. 532–554. Springer-Verlag, Berlin and New York.
Long, T. J., and Haber, A. H. (1965). *Radiat. Bot.* **5,** 223–231.
Schwartz, D., and Bay, C. E. (1956). *Amer. Natur.* **90,** 323–327.
Seliger, H. H., and McElroy, W. D. (1965). "Light." Academic Press, New York.
Shaw, M., and Manocha, M. S. (1965). *Can. J. Bot.* **43,** 747–755.
Shaw, M., Bhattacharya, P. K., and Quick, W. A. (1965). *Can. J. Bot.* **43,** 729–746.
Swanson, C. P. (1957). "Cytology and Cytogenetics." Prentice-Hall, Englewood Cliffs, New Jersey.
Walne, P. L., and Haber, A. H. (1968). *Radiat. Bot.* **8,** 399–406.
Woolhouse, H. W. (1967). *Symp. Soc. Exp. Biol.* **21,** 179–214.

Chapter 8

Photodynamic Ation of Laser Light on Cells*

I. L. CAMERON, A. L. BURTON, and C. W. HIATT

* Supported by Morrison Trust Grant R-A-12.

I. Introduction

There are a number of natural and essential photochemical processes by which cellular molecules such as the photosynthetic pigments (chlorophylls, carotenoids) absorb visible light energy for use in such processes as photosynthesis, vision, phototaxis, phototropism, photoperiodism, and photomorphogenesis (Seliger and McElroy, 1965). In 1898 a student named Raab showed that a nonpigmented protozoan could be sensitized to visible light by introduction of an appropriate dye (acridine). When the dye and the cells were combined in the presence of sunlight the cells were killed in about 6 min. Raab showed that neither the dye by itself nor the light by itself was harmful to the cells. Shortly after these initial experiments, Raab (1900) discovered that oxygen was consumed in the photosensitized killing process. By 1905, Raab's professors (Jodlbauer and Tappeiner, 1904, 1905; Tappeiner and Jodlbauer, 1904) had extended the concept that dye-sensitized photooxidation of biological systems was a general phenomenon, and they termed this reaction "the photodynamic phenomenon" or "photodynamic action." The early information on photodynamic action has been reviewed in an excellent monograph by Blum (1964). More recent reviews have been published by Spikes and Glad (1964), by Giese (1964), by Seliger and McElroy (1965), by Spikes (1968), and by Spikes and Livingston (1969). The Fourth International Jenaer Symposium in 1967 was devoted exclusively to this subject.

The present chapter is not intended to be a comprehensive review of the subject of photodynamic action but rather to introduce the reader to the essential characteristics of the phenomenon, especially as related to the use of laser light sources. Considerable information has accumulated on the biological and medical aspects of laser irradiation in both the visible and infrared wavelength ranges (see Whipple, 1965; Goldman, 1970). By combining the special properties of coherent light energy with the inherent specificity of the photodynamic process it is possible to effect the highly selective destruction of tissues, cells, and cell organelles.

II. Résumé of Some Characteristics and Guidelines Relating to Photodynamic Action

Photodynamic action may be defined as the sensitization of a biological system to visible light by a substance (a dye or pigment) that serves as a light absorber and brings about a destructive photochemical reaction in which molecular oxygen is consumed. Thus by definition, photodynamic reactions occur only in the presence of molecular oxygen. It is possible to prevent photodynamic action in some biological systems by the introduction of appropriate reducing agents. The absorption of a quantum of light is not dependent on temperature; thus the initial reactions resulting from the absorption may be temperature independent. Photodynamic action is an irreversible photochemical reaction in which the extent of the reaction is proportional to the quantity of light absorbed. Because of this fact it is possible to produce the same amount of photodynamic action using a lower intensity light with longer duration of exposure as can be obtained with higher intensity over a shorter duration of time. Because photodynamic reactions are essentially irreversible the effects or products of the reaction accumulate in the presence of continuous or of intermittent light. This irreversibility of the reaction is unlike the UV-induced cellular radiation damage which can be photoreactivated by visible light.

In order to produce photodynamic effects the light energy must be absorbed by the photosensitizer. Not all dyes are photosensitizers. Many cannot be induced to undergo the transition to an excited (triplet) state and do not, therefore, serve as photosensitizers. Agents that can sensitize biological systems to photodynamic action must become fixed to substrate molecules if they are to have an effect. A photodynamic sensitizer is not used up in the photoreaction and is, therefore, comparable to a catalyst. The site of action of many photosensitizing dyes is thought to be at the surface of molecules. However, it has also been reported that some photosensitizing dyes such as Acridine Orange can intercalate into deoxyribonucleic acid and photosensitize the nucleic acids to light. Clearly, then, preventing the uptake of a sensitizing dye on the part of the cell or other biological material can inhibit the photodynamic action. Conversely, shifting the ionic conditions, pH, etc., or changing the environment to increase the association of the dye with the biological materials will facilitate photodynamic action.

The photoreactions that are directly caused by ultraviolet irradiation are independent of molecular oxygen and on this basis are not considered true photodynamic phenomena. The molecular lesions caused by photosensitized oxidation have been shown to occur either as an oxidation of certain amino acids in proteins, i.e., tyrosine, tryptophan, methionine, and histidine molecules, and/or by cleavage of the purine nucleotides in nucleic acids, specifically, guanosine (Spikes and Livingston, 1969). If the reaction damages the nucleic acid, for example, the DNA of a virus, one anticipates a destructive response that is measured by testing the replication ability of the virus particles. In other cases, one might expect the protein molecules of a living cell to be directly damaged, leading to a change in structure and function of the cellular proteins. Thus, one might expect rapid killing of cells because of damage to essential proteins, and in other cases one might expect longer term damage as revealed by mutagenic or other destructive effects taking place at the level of DNA.

III. Examples of Photosensitizing Agents

Almost all photosensitizers are known to be fluorescent either when they are absorbed on a surface or when they are in solution. In relation to the sensitizing dye concentration, photodynamic action has a zero-order reaction rate, where the total number of reacting molecules is directly proportional to the total quanta of absorbed light. As pointed out above, the sensitizing molecule is not used up in the reaction and acts as a catalyst. Table I gives a list of sensitizing agents that have been shown to produce photodynamic action. This table was assembled from data found in Seliger and McElroy (1965) and in an unpublished bibliographic table compiled by Reich (1966). Not only do the dyes listed in Table I exhibit photodynamic action but most of these dyes also exhibit light-mediated carcinogenic activity when applied to the skin of mammals. Although porphyrin normally has a pronounced photodynamic action, it does not have this action in combination with hemes; that is to say when it contains metals such as magnesium or zinc, and when it is an integral part of the cell structure.

When a sensitizing agent is excited by light it can return to the unexcited ground state in different ways. The excitation energy can be (1) passed to the surroundings as heat, (2) lost by fluorescence, (3) used in a chemical transformation, or (4) transferred to another molecule. Cytochrome c can absorb visible light; however, during photosynthesis excitation energy is effectively channeled away so that harmful photodynamic action does not readily occur. It also seems likely that the carotenoid pigments found in the grana of chloroplasts serve as optical "filters" that protect or shield cytochrome c from visible light. A similar explanation may hold for other biological molecules that might be expected to photosensitize cells or organelles to visible light but do not seem to do so *in vivo*.

The various compounds in Table I absorb light at different wavelengths, each one having a characteristic absorption spectrum. Included among the list of photosensitizing agents in Table I is a wide variety of

TABLE I
Examples of Photosensitizing Agents

Acetylaminofluorene	Hematoporphyrin
Acridine orange	Hypericin
Acriflavine	Janus Green B
Allylarsonic acid	Khellin
Anthrasol	2′-Methyl-4-aminostrilbene
Aureomycin	Methylarsonic acid
Azure I and II	5-Methyl-1,2-benzanthracene
1,2-Benzanthracene	20-Methylcholanthrene
3,4-Benzpyrene	2′-Methyl-4-dimethylaminostilbene
Carbon tetrachloride	Methylene Blue
Chlorophyll	Methyl Orange
Chlorothiazide	Mg Pthalocyanine
Chlorpromazine	Neutral Red
Chloroform	New Methylene Blue N
Cholanthyrene	*p*-Aminobenzoic acid
Coal tar	Phenosafranine
Coproporphyrin	Phenothiazines
Cresyl Violet	Phloxine-B
Crystal Violet	Proflavine
Demethylchlortetracycline	Promazine
1,2,5,6-Dibenzanthracene	Protoporphyrin
Diethylstilbestrol	Quinacrine (atabrine)
p-Dimethylaminoazobenzene	Riboflavin
9,10-Dimethyl-1,2-dibenzanthracene	Rose Bengal
Eosin	Safranine
Erythrosin	Sulfonamides
Estrone	Sulfonylurea
Fluorescein	Thioflavine
5-Fluorouracil	Thionine
Furocoumarins	Toluidine Blue
Griseofulvin	Uroporphyrin

familiar and commonly used agents including antibiotics, hormones, porphyrin derivatives, dyes, and stains plus an antimalarial drug, a vitamin, and a cancer chemotherapeutic agent. A number of compounds on this list fall into the category of vital or supravital dyes that are used to stain living cells or tissues. These include Methylene Blue, Cresyl Violet, Neutral Red, Janus Green B, Acridine Orange, and Rose Bengal. The absorption spectrum of each of these vital dyes is illustrated in Fig. 1. In general, the vital dyes and many other compounds on the list can be added to biological systems without causing immediate or significant change or damage to the biological system, and these compounds with relatively little endogenous toxicity can be selected for use when one wants to study a cellular photodynamic response uncomplicated by toxic side reactions.

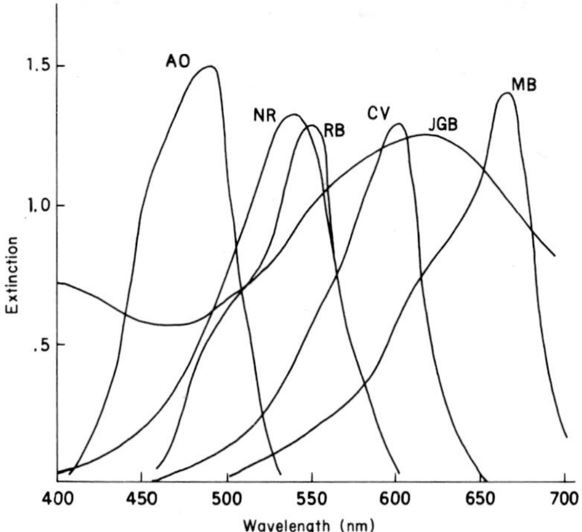

Fig. 1. Absorption curves of six vital dyes of different colors which are known to act as photosensitizers. The concentrations of the dyes was adjusted so that the extinction value at the peak of absorption is about the same in each case. From left to right the individual dyes are AO = Acridine Orange (an acridine dye), NR = Neutral Red (an azine dye), RB = Rose Bengal (a xanthene dye), CV = Cresyl Violet (a quinone-imine dye), JGB = Janus Green B (an azo dye), and MB = Methylene Blue (athiazin dye). (Data obtained from Lillie, 1969, and from Kasten, 1967.)

IV. Matching the Absorption Spectrum of the Photosensitizing Agent with the Wavelength of the Light Source

Photodynamic action depends upon the ability of the sensitizing dye to absorb a quantum of light that can then bring about a chemical reaction causing the photodynamic effect. Clearly, the sensitizing molecule can capture a quantum only in its absorption range. Thus, light energy delivered outside the absorption range will have no excitation effect. The wavelength of light that is effective in producing the photodynamic response must therefore be in the absorption range of the sensitizing agent to bring about a photodynamic response. In designing procedures to produce a photodynamic effect, it is essential that one be aware of the absorption spectra of the photodynamic-sensitizing substance and the wavelength of the incident light to be applied to the system. For example, Fig. 1 shows the absorption spectrum of a number of dyes including Methylene Blue (curve MB). Based on the absorption spectrum of Methylene Blue, one would expect maximum photodynamic activity to occur between the wavelengths of 600 to 700 nm. It should be apparent that by judicious choice of a specific and selective sensitizer dye in combination with a specific monochromatic light source, one can selectively damage specific biological structures or alter specific functions.

V. Comparison of Some Properties of Lasers and Other Light Sources

Laser light differs from ordinary light in several characteristics. Its waves are coordinated in both space and time. Laser beams are intense, chromatically pure (highly monochromatic), and directional in nature. Although specific lasers can be obtained that give light at specific wavelengths over a relatively large part of the electromagnetic spectrum, we are concerned primarily with light in the visible range and will, therefore, confine our discussion to lasers that produce visible light.

Table II lists some currently available lasers. As indicated, they can be used to produce either continuous light or short bursts of light known as pulses. The active material that is excited to produce the laser light, the emission spectrum or wavelength, the approximate power, and the approximate cost of the laser are indicated. For instance, one can see that a helium–neon (He–Ne) laser operates in a continuous manner at a 633 nm wavelength, and produces a beam with 1 to 50 mW of power. This laser beam is highly monochromatic at 633 nm with relatively high intensity. The beam that emerges from this instrument is about 2 mm in diameter and has considerable penetrating power for most living tissues. Comparison of the 633 nm emission spectrum as listed for the He–Ne

TABLE II
Lasers Currently Available

Kind of operation	Active material	Emission spectrum (wavelength in nm)	Power	Approximate cost ($, 1970)
Continuous	He–Cd	Blue (442)	50 mW	7,500
Continuous	He–Ne	Red (633)	1–50 mW	300–5,500
Pulsed up to 60 Cps	Ar	Blue, green, violet; main wavelengths 488 and 515	~1–2 mW	1,900 12,000
Continuous	Xe	Blue, green	~300–500 mW	3,000–6,000
Continouous	K	Blue, green, red, yellow	~500 mW	3,000–6,000
Pulsed	YAG:Nd YAG:Eu	1060 doubles to green	1 kW	4,000–20,000
Pulsed	Ruby	694 red	10 nsec 150 mW or 100 msec 10 joules	12,000

laser in Table II with the absorption spectra of the dyes in Fig. 1 indicates that the dyes Acridine Orange, Neutral Red, and Rose Bengal cannot be used as photosensitizer dyes with the He–Ne laser, but that Methylene Blue and Janus Green B would be expected to act as photosensitizers for light emission at 633 nm.

VI. Examples of Selective Destruction of Tissues, Cells, or Cell Organelles by Photodynamic Action Using Laser Light

A. Effects on Cells

Much work has been done on the photodynamic effects of light on living cells. The photodynamic effects that can be caused by laser light would not be expected to differ from those produced by ordinary light except where the monochromatic nature of the laser beam or the increased penetration or higher intensity might be involved. In this laboratory we have performed experiments to demonstrate the effects that a laser light beam has on the colorless protozoan ciliate *Tetrahymena pyriformis*. This organism can be grown in the presence of a rather intense He–Ne laser light at an intensity of 0.25 mW/2 mm^2 (Cameron and Burton, 1969). These cells will grow and reproduce in the presence of the continuous He–Ne red laser beam. In subdued light or darkness, *Tetrahymena* can also be cultured in the presence of Methylene Blue at a concentration of $1 \times 10^{-6}\,M$, without losing the ability to grow and reproduce. When one shines the red laser beam onto cells that have been photosensitized with Methylene Blue, one observes an almost immediate increase in cell swimming rate, which then begins to decrease after about 15 to 30 sec. This rate then gradually slows until death occurs after 3 to 9 min. When conducting these experiments it is important to use slides made from plastic with loose cover slips so as to assure an adequate oxygen supply to the solution. If glass slides and cover slips are used, and the edges of the cover slips are sealed with paraffin oil, a relatively high concentration of cells will rapidly deplete the oxygen tension of the media to negligible amounts. Under such conditions the laser beam has no apparent effect on the cell swimming or death rate. It should be noted that *Tetrahymena* can survive for a considerable time in an anaerobic environment without an appreciable change in swimming rate. By performing a series of experiments of this type we have been able to confirm the necessity of the presence of molecular oxygen to obtain a photodynamic effect in this system. (The above observations have been recorded on a sound motion picture film available from the authors on request.)

The beam from our He–Ne laser is approximately 2 mm in diameter. If this beam is projected through a shallow culture of cells, it can be seen that cells that chance to pass through the beam quickly die; those that do not enter the beam continue to survive. Because of the rapid swimming rate of *Tetrahymena,* all of the cells eventually come to die and concentrate within the laser beam. Thus, we have been able to use this procedure to concentrate the carcasses of cells. We believe that the method can be adapted for use in determining swimming rates of microorganisms in solution simply by enumerating the number of cells either remaining outside the beam or those that are concentrated in the beam.

A report by Rounds *et al.* (1968) suggests that laser light may be useful in the management of malaria. In this study erythrocytes that were infected by the malarial parasite were treated with Methylene Blue and exposed to an intense ruby laser beam. Those erythrocytes that were parasitized by malaria demonstrated a wavelength-specific photolysis. The authors demonstrated the same effect on the parasitized erythrocytes from ducks as well as those from man. When the duck erythrocytes were sensitized by Methylene Blue and exposed to the laser beam, 100% of the parasitized cells were selectively lysed. Curiously enough, there was also a marked reduction of infected erythrocytes (about a 20% decrease) in samples of malarial duck blood that were not treated with dye. Rounds *et al.* believe that this is caused by an endogenous sensitizer derived from the breakdown of hemoglobin. Rounds and his group suggest that it may be possible to destroy infected erythrocytes in centrifuged whole blood without harming the white blood cells.

B. Microbeams through a Microscope

The use of laser microbeams to perform cell surgery has recently come to the forefront. One of the first applications of laser microirradiation of living cells was demonstrated by Bessis and Ter-Pogossian in 1965. In these studies a ruby laser beam was concentrated from a 5 mm spot to a 2.5 μ spot by passing it through the ocular and objectives of a phase microscope. This beam caused the selective destruction of mitochondria. A series of recent reports on cell surgery by laser microbeams has come from The Laser Biology Laboratory of The Pasadena Foundation for Medical Research (Berns *et al.*, 1969a,b, 1970; Berns and Rounds, 1970a,b; Rounds *et al.*, 1968). A laser microbeam apparatus devised by this group is diagrammed in Fig. 2. In this system an argon ion laser

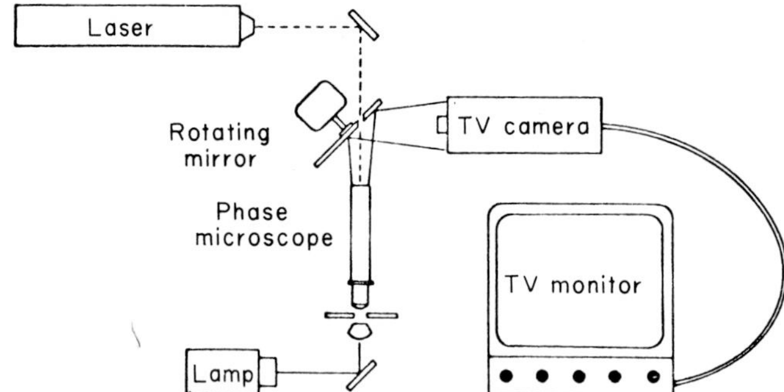

Fig. 2. Laser microbeam apparatus used by Berns and Rounds (1970a,b) to selectively destroy cell organelles (see text for further explanation). (The figure is reproduced with the permission of the editors of The Annals of the New York Academy of Sciences.)

beam with principal wavelengths of 515 and 488 nm with a power output of 1 W was pulsed repetitively at 60 pulses/sec. A microbeam of between 0.5 and 0.75 μ can be produced with such a system simply by aligning and passing the beam through the optical system of the microscope. As shown in Fig. 2, the laser beam is reflected down through a circle in a rotating mirror, the speed of which is synchronized with the laser pulse frequency. A trigger is located on the back surface of the mirror so that each pulse from the laser passes through the aperture of the rotating mirror. The light from a microscope lamp is equipped with a filter to screen out extraneous light in the same range as that which would affect the sensitized cells. The filtered light from the microscope lamp is focused through the specimen against the rotating mirror and into a TV camera, so that all events that occur in the cell can be continuously monitored. Any desired target within a cell can be adjusted to the focal point of the laser by means of a mechanical stage and a cross-hair sight. This apparatus has been used to produce local lesions in chromosomes, nucleoli, and mitochondria.

In the case of chromosomes, Acridine Orange in concentrations ranging from 2.5×10^{-3} to $2.2 \times 10^{-6}\%$ was used to sensitize the chromosomes. Acridine Orange intercalates into the structure of DNA. By use of the argon laser microbeam a lesion as small as 6 μ can be placed in preselected sites of the chromosomes. Their studies show the selective destruction of the DNA in specifically irradiated chromosome regions. Such chromosomal lesions did not stop cell division, cytokinesis, or the reconstruction of nuclear membranes and nucleoli. Indeed, the cells were shown to survive such microbeam irradiation and to stay viable for several days. Berns *et al.* (1969a,b) point out that this technique can be used to study chromosome structure and repair, the process of cell division, and to assist in the genetic mapping of individual chromosomes. Studies have also been performed to destroy selectively the nucleoli of cells, thus interfering with ribosome biogenesis and ultimately blocking protein synthesis. The antimalarial drug quinacrine hydrochloride was used as the photosensitizer. Again, the argon laser microbeam apparatus was used. After the nucleolus or nucleoli have been irradiated and the quinacrine has been washed from the cells, the cells were then labeled with radioactive uridine, a precursor of RNA. After 30 min the cells were fixed and prepared for radioautography. RNA synthesis was drastically reduced in cells that had received microbeam irradiation of the nucleoli. Control lesions in other regions of the nucleus did not produce significant alteration in RNA synthesis.

In earlier microbeam studies mitochondria were sensitized specifically with Janus Green B, but Berns and Rounds (1970a,b) and Berns *et al.* (1970), using an intense argon laser beam, found that no photosensitizing agent was actually necessary if the intensity of the laser was high enough. Their studies show that individual mitochondria can be destroyed selectively within a cardiac muscle cell. Cardiac muscle cells normally pulsate (contract and relax) in culture. Microbeam irradiation of a single mitochondrion did not adversely affect the cell contractions. However, if several mitochondria were irradiated, the contractions ceased and cell death occurred. The cytochromes ($c + c_1$) absorb some light in the blue–green region of the spectrum and may act as endogenous photosensitizers in mitochondria.

C. Bloodless Surgery

We recently performed some feasibility studies to test the potential usefulness of laser light and photodynamic action in performing what may be called bloodless surgery. We isolated the sciatic nerve on each side of an anesthetized rabbit. One of the nerves was injected with Methylene Blue (10^{-6} M) and irradiated for 1 min with a He–Ne laser light at an intensity of 0.25 mW/2 mm^2. The other nerve was exposed and also irradiated for 1 min but without prior injection of Methylene Blue. Previous studies have indicated that Methylene Blue has no adverse dark effect. The animal was sacrificed after 2 days, and the treated regions of the sciatic nerves were excised and fixed for histological examination. Figure 3 shows a longitudinal section of the nerve that was irradiated by the laser beam without Methylene Blue. The nerve appears normal in all respects. Figure 4 shows a similar longitudinal section of the nerve that

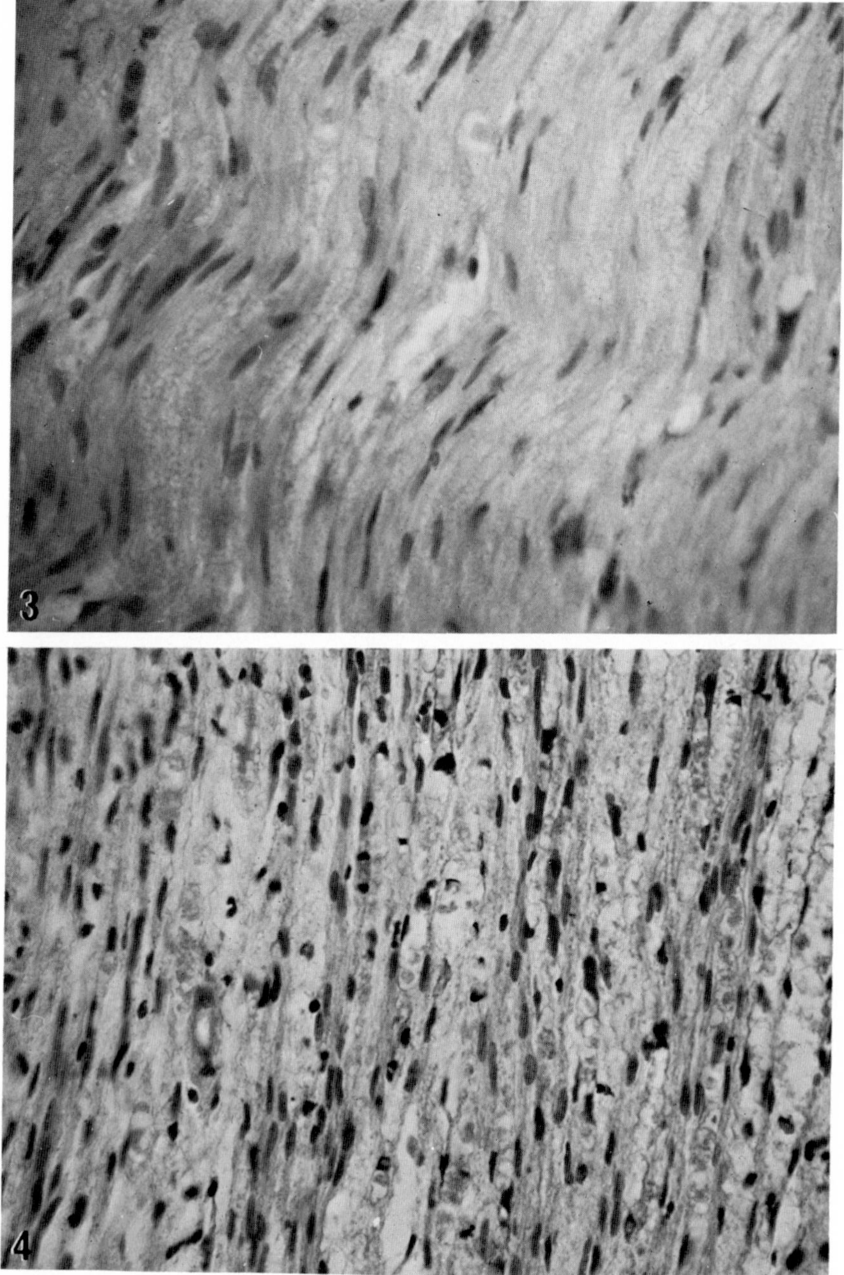

Figs. 3 and 4. Photomicrographs (same magnification) of longitudinal histological sections of the sciatic nerves of a rabbit. The one sciatic nerve shown in Fig. 3

had received Methylene Blue prior to laser irradiation. Vacuolation of cells is apparent as well as pyknotic cell nuclei. An increased number of mononuclear cells was also seen. It was apparent that the Methylene Blue in combination with laser irradiation had caused severe damage to this nervous tissue. The work of Jungstandt and Berg (1967) has already demonstrated that after sensitization with Methylene Blue or thiopyronin, transplantable intradermal tumors in mice could be eliminated by light treatment. Clearly, then, the potential for selective use of photodynamic action in tumor destruction has been demonstrated, and the feasibility of specific attack upon organelles suggests many interesting applications.

The tumor studies of Jungstandt and Berg (1967) were performed with polychromatic light. Comparable studies with laser beams might demonstrate deeper penetration of tissues with faster and more localized results. By means of narrow gage syringes appropriate vital photosensitizing dyes could be injected into the area of a tumor, or, following the technique of Jungstandt and Berg, dimethylsulfoxide might be employed to facilitate penetration of the dye into the surrounding tissue. The laser beam could then be focused through the overlying skin or tissue to the site of the tumor. Only that tissue that would receive both the dye and the light irradiation would be damaged. Thus, tumors lying deep within the body could be selectively destroyed without damaging overlying structures. As an adjunct we also propose that fiber optic systems might be used to selectively irradiate tissues not normally accessible to light, i.e., the stomach, intestine, urethra, urinary bladder, uterus, etc.

References

Berns, M. W., and Rounds, D. E. (1970a). *Ann. N. Y. Acad. Sci.* **168,** 550.

Berns, M. W., and Rounds, D. E. (1970b). *Sci. Amer.* **222,** 99.

Berns, M. W., Olson, R. S., and Rounds, D. E. (1969a). *J. Cell Biol.* **43,** 621.

Berns, M. W., Rounds, D. E., and Olson, R. S. (1969b). *Exp. Cell Res.* **56,** 292.

Berns, M. W., Gamaleja, N., Olson, R., Duffy, C., and Rounds, D. E. (1970). *J. Cell. Physiol.* **76,** 207.

Bessis, M., and Ter-Pogossian, M. M. (1965). *Ann. N. Y. Acad. Sci.* **122,** 689.

Blum, H. F. (1964). "Photodynamic Action and Diseases Caused by Light." Hafner, New York.

Cameron, I. L., and Burton, A. (1969). *J. Cell Biol.* **43,** 165a.

---

received irradiation from a He–Ne laser for 1 min while the rabbit's other sciatic nerve, shown in Fig. 4, received an injection of Methylene Blue dye, then a 1-min exposure to laser irradiation. The animal was sacrificed 2 days after laser treatment. The nerve in Fig. 3 is normal in histological appearance while the nerve in Fig. 4 shows massive destruction (see text for more details).

Giese, A. C. (1964). *In* "Photophysiology" (A. C. Giese, ed.), Vol. 1, pp. 4–6. Academic Press, New York.

Goldman, L. (1970). *Ann. N. Y. Acad. Sci.* **168.**

Jenaer Symposium IV. (1967). *Stud. Biophys.* **3.**

Jodlbauer, A., and Tappeiner, H. (1904). *Muenchen. Med. Wochenschr.* **26,** 1139.

Jodlbauer, A., and Tappeiner, H. (1905). *Deut. Arch. Klin. Med.* **82,** 520.

Jungstand, W., and Berg, H. (1967). *Stud. Biophys.* **3,** 225.

Kasten, F. H. (1967). *Int. Rev. Cytol.* **21,** 141–202.

Lillie, R. D. (1969). "H. J. Conn's Biological Stains," 8th ed. Williams & Wilkins, Baltimore, Maryland.

Raab, O. (1900). *Z. Biol.* (*Munich*) **39,** 524.

Reich, A. K. (1966). "Bibliography on Photodynamic Action as Related to Biological Products." NIH, Bethesda, Maryland.

Rounds, D. E., Opel, W., Olson, R. S., and Sherman, I. W. (1968). *J. Cell Biol.* **32,** 616.

Seliger, H. H., and McElroy, W. D. (1965). "Light." Academic Press, New York.

Spikes, J. D. (1968). *In* "Photophysiology" (A. C. Giese, ed.), Vol. 3, pp. 33–64. Academic Press, New York.

Spikes, J. D., and Glad, B. W. (1964). *Photochem. Photobiol.* **3,** 471.

Spikes, J. D., and Livingston, R. (1969). *Advan. Radiat. Biol.* **3,** 29–121.

Tappeiner, H., and Jodlbauer, A. (1904). *Deut. Arch. Klin. Med.* **80,** 427.

Whipple, H. E. (1965). *Ann. N. Y. Acad. Sci.* **122.**

Author Index

Numbers in italics refer to the pages on which the complete references are listed.

C

H

I

J

T

Z

Subject Index

D

E

F

G

H

I

L

M